T0282312

LONDON MATHEMATICAL SOCIETY LECTURE !

Managing Editor: Professor J.W.S. Cassels, Department of Pure Mathematics and Mathematical Statistics, University of Cambridge, 16 Mill Lane, Cambridge CB2 1SB, England

The books in the series listed below are available from booksellers, or, in case of difficulty, from Cambridge University Press.

London Mathematical Society Lecture Note Series. 188

Local Analysis for the Odd Order Theorem

Helmut Bender
Universität Kiel

and

George Glauberman
University of Chicago

with the assistance of
Walter Carlip
Ohio University

CAMBRIDGE
UNIVERSITY PRESS

Published by the Press Syndicate of the University of Cambridge
The Pitt Building, Trumpington Street, Cambridge CB2 1RP
40 West 20th Street, New York, NY 10011, USA
10, Stamford Road, Oakleigh, Melbourne 3166, Australia

© Cambridge University Press 1994

First published 1994

Library of Congress cataloging-in-publication data available

British Library cataloguing in publication data available

ISBN   0 521 45716 5   paperback

Transferred to digital printing 2005

In memory of R. H. Bruck

Contents

Contents

Preface

About 30 years ago, Walter Feit and John G. Thompson [8] proved the Odd Order Theorem, which states that all finite groups of odd order are solvable. In the words of Daniel Gorenstein [15, p. 14], "it is not possible to overemphasize the importance of the Feit-Thompson Theorem for simple group theory." Their proof consists of a set of preliminary results followed by three parts–local analysis, characters, and generators and relations–corresponding to Chapters IV, V, and VI of their paper (denoted by **FT** here). Local analysis of a finite group G means the study of the structure of, and the interaction between, the centralizers and normalizers of nonidentity p-subgroups of G. Here Sylow's Theorem is the first main tool. The main purpose of this book is to present a new version of the local analysis of a minimal counterexample G to the Feit-Thompson Theorem, that is, of Chapter IV and its preliminaries. We also include a remarkably short and elegant revision of Chapter VI by Thomas Peterfalvi in Appendix C.

What we would ideally like to prove, but cannot, is that each maximal subgroup M of G has a nonidentity proper normal subgroup M_0 such that

(1) $C_{M_0}(a) = 1$, for all elements $a \in M - M_0$,
(2) $M_0 \cap M_0{}^g = 1$, for all elements $g \in G - M$,
(3) M_0 is nilpotent,
(4) M/M_0 is cyclic,

and such that the totality of these subgroups M_0, with M ranging over all of the maximal subgroups of G, forms a partition of G:

(5) each nonidentity element of G lies in exactly one of the subgroups M_0.

Relating each step in our procedure (as well as the main results, given in Section 16) to this hypothetical goal will help give the reader a sense of direction and motivation: after the normal Hall subgroup M_σ has been introduced in Section 10, it can be read as M_0. (Section 16 is self-contained, except for notation from Section 1, and can be read as a supplement to this introduction.)

In addition, we strongly recommend first studying a theorem of Feit, Thompson, and Marshall Hall [7], the immediate predecessor of **FT**, which proved solvability under the additional CN-condition: the centralizer of every nonidentity element of G is nilpotent. The local analysis part of its proof leads to conditions (1)–(5) for a minimal counterexample G. A guide to reading this miniature model for **FT** and our work is given in Appendix D. This theorem is actually needed in **FT** [8, p. 983], although not for the part covered by this book. Incidently, the conditions (1)–(5) above clearly imply the CN-condition. Furthermore, (1) means that M is a Frobenius group with kernel M_0, and thus implies (3) by a very special case of a theorem of Thompson (Theorem 3.7).

The Odd Order Theorem was originally conjectured in the nineteenth century. The first essential step toward its proof was taken by Michio Suzuki [25] in 1957. He showed that CA-groups of odd order are solvable; here CA means that all centralizers are abelian. In this case it is a routine matter to derive (1)–(5), with all M_0 abelian. Suzuki's contribution, a model for the later CN-paper, was mainly character-theoretic. Conditions (1)–(2) and variations thereof occur in much more general situations as the end result of local analysis, and it is therefore of fundamental importance for finite group theory that they have strong character theoretic implications. See [14, pp. 139-148], [17, pp. 195-205], or [26, pp. 281-294] for details.

It is the purpose of this book to make the Feit-Thompson Theorem more accessible to a reader familiar with some standard topics in finite group theory, such as Chapters 1–8 of Gorenstein's first book [14] (henceforth denoted by **G**). However it is possible to manage comfortably with considerably less reading. We give information about prerequisites in Appendix A. For the convenience of the reader, strictly necessary references to other works appear only in Chapter I, and refer only to **G**. Further information about the influence of the theorem and its proof, together with a detailed description of the proof, may be found in **G**, pp. 450-461, and in [15, pp. 13-39].

As stated above, our main text and Appendix C correspond to Chapters IV and VI of **FT** and the necessary preliminaries. As to the missing link, the necessary character theory, we must refer the reader to Chapter V of **FT** or to some unpublished work of David Sibley, who has obtained very interesting improvements [23, pp. 385-388]. Fortunately, Chapter V of the original paper is somewhat less complicated than Chapter IV.

We hope that in the not too distant future there will be a unified revised proof of the Feit-Thompson Theorem. In addition, we and others have some thoughts now for further improving this work; in this spirit, we include a few results that are not needed for Chapter V of **FT** or for Sibley's work. However, in view of the considerable interest expressed in this work and the improvements and corrections sent to us by readers of preliminary versions, we have decided to publish the work now as a set of lecture notes.

In a sense, the first steps toward the writing of this book were taken in 1962, when the second author began to study a preprint of the Odd Order Paper, with the encouragement and assistance of his Ph. D. advisor, R. H. Bruck. However, the actual writing of a revision started with a class at the University of Chicago in the Winter and Spring Quarters of 1975.

We wish to thank the members of the 1975 class (particularly David Burry, Noboru Itô, Richard Niles, David T. Price and Jeffrey D. Smith) and of a similar class given in Winter, 1986 (particularly Curtis Bennett, Walter Carlip, Diane Herrmann, Arunas Liulevicius, Peter Sin, and Wayne W. Wheeler). In addition, preliminary versions of this work were read by Paul Lescot, Thomas Peterfalvi, and David Sibley, and studied in seminars at the University of Florida and Wayne State University, led by László Héthelyi (of Technical University, Budapest) and by Daniel Frohardt, David Gluck and Kay Magaard, respectively. We thank each of these individuals and the members of these seminars for their corrections and suggestions.

For permission to include unpublished work, we thank David Sibley (Theorem 14.4, Corollary 15.9); I. Martin Isaacs (Appendix B); Walter Carlip and Wayne W. Wheeler (Appendix C); and especially Walter Feit and John G. Thompson (Theorem 15.8, Corollary 15.9, Appendix E). Appendix C is based on a beautiful revision [22] of Chapter VI of **FT**, for which we thank the author, Thomas Peterfalvi.

We are particularly indebted to Professors Feit and Thompson for their help and encouragement throughout the preparation of this work.

We note with great sadness the deaths of two individuals who also played instrumental roles: R. H. Bruck and Daniel Gorenstein. Without them this work might never have been started nor ever have been completed.

As this book has gone through many stages and vicissitudes in twenty years, there is a danger that we have inadvertently overlooked some individuals to whom thanks are due. To them we sincerely apologize.

During the preparation of parts of this work the second author enjoyed the support of the Guggenheim Foundation and the National Science Foundation, and the hospitality of the Mathematical Institute, Oxford; Jesus College, Oxford; Kansas State University; and Universität Kiel. He thanks each of these institutions. He also thanks the members of his family for their helpful patience, forbearance, or nagging.

An earlier, complete version of this work was prepared by the second author with the assistance of Alexandre Turull in 1979. The present version was prepared with the assistance of Walter Carlip. Both have made valuable corrections and improvements in the mathematical content and the wording of the texts, particularly Dr. Carlip, who has also worked assiduously, over the course of many years to put preliminary drafts into TEX and to produce the final camera-ready copy printed here. We thank both for their efforts.

CHAPTER I

Preliminary Results

Here we give general results about finite groups, mainly solvable groups and p-groups, including some special properties of groups of odd order. In Chapters II–IV we will apply the results of this section to a hypothetical minimal counterexample to the Odd Order Theorem. As mentioned in the preface, all necessary references in this chapter are taken from **G**.

1. Elementary Properties of Solvable Groups

Suppose G is a group. We say that a group A *operates* on G, or A is an *operator group* on G, if there is given a homomorphism ϕ from A into Aut G. In this case we usually write x^α instead of $\phi(\alpha)(x)$ for $x \in G$ and $\alpha \in A$. We say that A *fixes* an element x of G, or that x is A-*invariant*, if $x^\alpha = x$ for every $\alpha \in A$. We say that A *fixes* a nonempty subset S of G, or that S is A-*invariant*, if each element of A fixes S as a set. As in **G**, pp. 30, 33, the set (group) of all A-invariant elements of G will be denoted by $C_G(A)$. Similarly, if S is a nonempty subset of G, $C_A(S)$ will denote the set of all elements of A that fix every element of S.

We will frequently use the fact (**G**, p. 18) that if H and K are subgroups of a group G, then

$$[H, K] \lhd \langle H, K \rangle.$$

By applying this fact to the semidirect product of a group G by an operator group A, we see that $[G, A]$ is a normal subgroup of G fixed by A. As in **G**, p. 19, $[G, A, A]$ will denote $[[G, A], A]$. Also, we say A *stabilizes* a normal series

$$G = G_0 \supseteq G_1 \supseteq \cdots \supseteq G_n = 1$$

of G if each G_i is A-invariant and A acts trivially on each factor G_{i-1}/G_i, $1 \leq i \leq n$.

1

Suppose that A is an operator group on a group G. As in **FT**, p. 840, we say that A acts in a *prime manner* on G if

$$C_G(\alpha) = C_G(A) \text{ for all } \alpha \in A^{\#}.$$

(Note that this must occur if $|A|$ is prime and that we allow $A = 1$.) We say that A acts *regularly*, or in a *regular manner* on G if

$$C_G(\alpha) = 1 \text{ for all } \alpha \in A^{\#}.$$

(Thus, if A acts regularly, then $A \subseteq \operatorname{Aut} G$ and A acts in a prime manner on G. This disagrees slightly with the definition in **G**, p. 39, which requires also that $A \neq 1$.)

In the subsequent text we will write $H \lhd\lhd G$ to mean that H is a *subnormal* subgroup of G. This means that H is a member of a normal series of G (**G**, Exercise 1.5, p. 13). Equivalently, there exists a series

$$H = H_0 \lhd H_1 \lhd \cdots \lhd H_n = G.$$

We use the property that every subgroup of a nilpotent group G is subnormal in G. This follows immediately from the fact that proper subgroups of a nilpotent group are properly contained in their normalizers (**G**, Theorem 2.3.4, p. 22).

All groups considered in this work will be finite except when explicitly stated otherwise.

For later use we make the following definition. Given a prime p and a group G, we say that G has *p-length one* if $G = \mathcal{O}_{p',p,p'}(G)$. (This differs slightly from the definition in **G**, p. 227, in that our definition includes groups of p'-order, that is, groups that, according to the usual definition, would have p-length zero.)

A group G is called a *Z-group* if all of its Sylow subgroups are cyclic.

For any subset T of G we define

$$\mathscr{C}_G(T) = \{\, t^g \mid t \in T \text{ and } g \in G \,\}.$$

A nonempty subset X of G is a *TI-subset* of G if $X \cap X^g \subseteq 1$ for all $x \in G - N(X)$. In particular, a nonidentity proper subgroup H of G is a *TI-subgroup* of G if $H \cap H^g = 1$ for all $g \in G - N(H)$.

In the text that follows we will denote by $\mathcal{E}_p(G)$ the set of all elementary abelian p-subgroups of G; $\mathcal{E}_p{}^*(G)$ the set of all maximal elementary abelian p-subgroups of G; and $\mathcal{E}_p{}^i(G)$ the set of all elementary abelian subgroups of order p^i in G (where i is a positive integer). We let $\mathcal{E}(G)$ be the union of the sets $\mathcal{E}_q(G)$ for all primes q. We define $\mathcal{E}^*(G)$ and $\mathcal{E}^i(G)$ analogously.

For a prime p, a p-group R will be called *narrow* if it contains no elementary abelian subgroup of order p^3 or if it contains a subgroup R_0 of order p and a cyclic subgroup R_1 such that $C_R(R_0) = R_0 \times R_1$. (This definition is not standard and is used only in this book. It corresponds to the definition of π^* on p. 845 of **FT**.)

Lemma 1.1. Suppose that M is a minimal normal subgroup of a finite group G. If M is solvable, then $M \subseteq Z(F(G))$ and is elementary abelian.

Proof. Elementary. $\square$

Proposition 1.2 (P. Hall). Suppose that G is a solvable group and that $G^* \lhd G$. Let $\mathscr{D}$ be the set of all chief factors U/V of G. Let $\mathscr{D}^*$ be the set of all chief factors U/V of G for which $U \subseteq F(G^*)$. Then

$$F(G^*) = \bigcap_{U/V \in \mathscr{D}} C_{G^*}(U/V) = \bigcap_{U/V \in \mathscr{D}^*} C_{G^*}(U/V).$$

Proof. Let

$$H = \bigcap_{U/V \in \mathscr{D}} C_{G^*}(U/V) \quad \text{and} \quad H^* = \bigcap_{U/V \in \mathscr{D}^*} C_{G^*}(U/V).$$

Take $U/V \in \mathscr{D}$. Then U/V is a minimal normal subgroup of G/V. By Lemma 1.1,

$$U/V \subseteq Z(F(G/V)).$$

Since $F(G^*)V/V$ is nilpotent and is also normal in G/V, we know that $F(G^*)V/V \subseteq F(G/V)$. Hence $F(G^*)V/V$ centralizes U/V. As U/V was taken arbitrarily, $F(G^*) \subseteq H$.

Clearly $H \subseteq H^*$. To complete the proof, we assume that $H^* \not\subseteq F(G^*)$ and obtain a contradiction. Let K be a normal subgroup of G minimal with respect to the property that $K \subseteq H^*$ and $K \not\subseteq F(G^*)$. Take a chief series for G that includes K, and let

$$(1.1) \qquad\qquad K = K_0 \supset K_1 \supset \cdots \supset K_n = 1$$

be the part of the chief series from K to 1. By the choice of K, we have $K_1 \subseteq F(G^*)$. Hence, for $i = 2, \ldots, n$, we have $K_{i-1}/K_i \in \mathscr{D}^*$ and, since $K \subseteq H^*$, we have $[K_{i-1}, K] \subseteq K_i$. Since K is solvable, K/K_1 is abelian and $[K_0, K] = [K, K] \subseteq K_1$. Thus the series (1.1) is a central series for K. Hence K is nilpotent. Therefore $K \subseteq F(G^*)$, a contradiction. $\square$

Proposition 1.3 (P. Hall). Suppose that G is a solvable group. Then $C_G(F(G)) \subseteq F(G)$.

Proof. Let $G^* = G$ in Proposition 1.2. $\square$

Proposition 1.4. Suppose that G is a solvable group, A is a group of automorphisms of G, and $(|A|, |G|) = 1$. Then A acts faithfully on $F(G)$.

Proof. We may assume that A is cyclic. Let X be the semidirect product of G by A. Then X is solvable. We embed A and G in X. Let $\sigma = \pi(A)$ and $F = F(X)$.

Since A is certainly a Hall σ-subgroup of X and $A\mathcal{O}_\sigma(F)$ is a σ-group, $A = A\mathcal{O}_\sigma(GF) \supseteq \mathcal{O}_\sigma(F)$. As $A \subseteq \operatorname{Aut} G$ and

$$[\mathcal{O}_\sigma(F), G] \subset \mathcal{O}_\sigma(F) \cap G = 1,$$

we have $\mathcal{O}_\sigma(F) = 1$. Thus

$$F = \mathcal{O}_\sigma(F) \times \mathcal{O}_{\sigma'}(F) = \mathcal{O}_{\sigma'}(F) \subseteq \mathcal{O}_{\sigma'}(X) = G.$$

Clearly $F = F(G)$. By Proposition 1.3,

$$C_A(F) = A \cap C_X(F(X)) \subseteq A \cap F(X) \subseteq A \cap G = 1. \quad \square$$

Proposition 1.5. Suppose that G is a solvable group, A is an operator group on G, and $(|A|, |G|) = 1$. Let π be a set of primes. Then:

  (a) A fixes some Hall π-subgroup of G;
  (b) every A-invariant π-subgroup of G is contained in an A-invariant Hall π-subgroup of G;
  (c) if H_1 and H_2 are A-invariant Hall π-subgroups of G, then H_1 and H_2 are conjugate by an element of $C_G(A)$;
  (d) if H is any A-invariant normal subgroup of G, then $C_{G/H}(A)$ is the image of $C_G(A)$ in G/H; and
  (e) if $C_G(A)$ contains a Hall π'-subgroup of G, then $[G, A] \subseteq \mathcal{O}_\pi(G)$.

Proof. Statements (a), (c), and (d) follow from P. Hall's theorem on solvable groups (**G**, Theorem 6.4.1, p. 231) and from the proof of Theorem 6.2.2, pp. 224–5 of **G**.

To prove (b) we proceed by induction on $|G|$. Let K be an A-invariant π-subgroup of G and M a minimal A-invariant normal subgroup of G. If G itself is a π-group, there is nothing to prove, and so we may assume G is not a π-group. Now KM/M is an A-invariant π-subgroup of G/M so, by induction, there exists an A-invariant Hall π-subgroup H/M of G/M that contains KM/M. Thus H is an A-invariant subgroup of G such that $K \subseteq H \subseteq G$ and $|H|_\pi = |G|_\pi$. If $H \neq G$, we can apply induction to H to conclude that K is contained in an A-invariant Hall π-subgroup of H and we are done. If $H = G$, then M is a normal Sylow p-subgroup of G for some prime p outside π. By (a), G has an A-invariant Hall π-subgroup Q and clearly $G = QM$ with $Q \cap M = 1$. Now $|Q \cap KM| = |K|$, and hence K and $Q \cap KM$ are both A-invariant Hall π-subgroups of KM. By (c), there exists an element $x \in C_{KM}(A)$ such that $K = (Q \cap KM)^x \subseteq Q^x$. Clearly Q^x is an A-invariant Hall π-subgroup of G.

To prove (e), let H be an A-invariant Hall π-subgroup and let K be a Hall π'-subgroup of G contained in $C_G(A)$. Then $G = KH$. Therefore

$$[G, A] = \left\langle h^{-1}k^{-1}k^\alpha h^\alpha \mid k \in K,\, h \in H,\, \alpha \in A \right\rangle \subseteq H.$$

Since $[G, A] \lhd G$, we have $[G, A] \subseteq \mathcal{O}_\pi(G)$. $\square$

Proposition 1.6. Suppose that G is a solvable group, A is an operator group on G, and $(|A|, |G|) = 1$. Then:

    (a) $G = C_G(A)[G, A] = [G, A]C_G(A)$;
    (b) $[G, A, A] = [G, A]$;
    (c) if $[G, A, A] = 1$, then A acts trivially on G;
    (d) if G is abelian, then $G = C_G(A) \times [G, A]$; and
    (e) if G is abelian and $C_G(A)$ contains every element of prime order in G, then A acts trivially on G.

Proof. For (a), let $H = [G, A]$ in Proposition 1.5(d). For (b) and (c), see the proof of **G**, Theorem 5.3.6, p. 181. For (d), see the proof of **G**, Theorem 5.2.3, p. 177. Finally, note that (e) follows from (d). $\square$

In the following lemma we list some of the basic properties of the Frattini subgroup of a finite group.

Lemma 1.7. Suppose that G is a group and R is a p-group for some prime p. Then:

    (a) if H is a subgroup of G and $G = H\Phi(G)$, then $G = H$;
    (b) $R/\Phi(R)$ is elementary abelian;
    (c) $\Phi(R) = 1$ if and only if R is elementary abelian; and
    (d) $\Phi(R) = \langle R', x^p \mid x \in R \rangle$.

Proof. (a) **G**, Theorem 5.1.1, p. 173. (b) **G**, Theorem 5.1.3, p. 174. (c) **G**, Theorem 5.1.3, p. 174. (d) Let $S = \langle R', x^p \mid x \in R \rangle$. By (b), $S \subseteq \Phi(R)$. Since R/S is elementary abelian and $\Phi(R/S) = \Phi(R)/S$, (c) yields (d). $\square$

Theorem 1.8 (Burnside). Suppose that A is an operator group on a p-group R and $(|A|, |R|) = 1$. Assume that A centralizes $R/\Phi(R)$. Then A centralizes R.

Proof. By Proposition 1.5(d), $R = C_R(A)\Phi(R)$. By Lemma 1.7(a), $R = C_R(A)$. (This is **G**, Theorem 5.1.4, p. 174.) $\square$

Lemma 1.9. Suppose that π is a set of primes, G is a finite solvable π-group, and A is an operator group on G that stabilizes a normal series of G. Then $A/C_A(G)$ is a π-group.

Proof. It suffices to show that A acts trivially on G if A is a π'-group. This follows from Proposition 1.5(d) by induction on the length of the normal series. $\square$

Proposition 1.10. Suppose that A is an operator group on a nilpotent group G and $(|A|, |G|) = 1$. Let $C = C_G(A)$. If $C_G(C) \subseteq C$, then A acts trivially on G.

Proof. Assume $C_G(C) \subseteq C$. Take $x \in N_G(C)$. For each $a \in A$ and $y \in C$, we know that $x^{-1}yx = (x^{-1}yx)^a = (x^a)^{-1}yx^a$ and $x^a x^{-1}$ centralizes y. Thus $x^a x^{-1} \in C_G(C) \subseteq C$. As x and a are arbitrary, A centralizes $N_G(C)/C$. Thus A stabilizes the normal series

$$N_G(C) \supseteq C \supseteq 1,$$

and hence, by Lemma 1.9, A acts trivially on $N_G(C)$. Thus $N_G(C) \subseteq C$. As G is nilpotent, $C = G$. Hence A acts trivially on G. □

Theorem 1.11. Suppose that p is an odd prime, G is a p-group, and A is a p'-group of operators on G that acts trivially on $\Omega_1(G)$. Then A acts trivially on G.

Proof. **G**, Theorem 5.3.10, p. 184. □

Corollary 1.12. Suppose that p is an odd prime, G is a p-group, E is an elementary abelian subgroup of G, and A is a p'-group of operators on G. Assume that A fixes every element of order p in $C_G(E)$. Then A acts trivially on G.

Proof. Let $C = C_G(A)$. Since $E \subseteq C_G(E)$, we know that $E \subseteq C$. Therefore $C_G(C) \subseteq C_G(E)$ and A fixes every element of $\Omega_1(C_G(C))$. Since p is odd, A fixes every element of $C_G(C)$ by Theorem 1.11. Consequently $C_G(C) \subseteq C$. By Proposition 1.10, A acts trivially on G. □

Theorem 1.13 (J. G. Thompson). Suppose that p is an odd prime and G is a nontrivial p-group. Then G contains a characteristic subgroup H that enjoys the following properties:
   (a) $[H, G] \subseteq Z(H)$;
   (b) H has nilpotence class at most two;
   (c) H has exponent p; and
   (d) $C_{\text{Aut } G}(H)$ is a p-group.

Proof. This follows from Thompson's Critical Subgroup Theorem (**G**, Theorem 5.3.11, p. 185). As in Theorem 5.3.13 of **G** (p. 186), we let C be a critical subgroup of G and examine the properties of $\Omega_1(C)$. Let $H = \Omega_1(C)$. Then (b), (c), and (d) are proven in **G** (Theorem 5.3.13, p.186). Since C is a critical subgroup of G, $[G, C] \subseteq Z(C)$. Thus

$$[G, H] = [G, \Omega_1(C)] \subseteq [G, C] \cap H \subseteq Z(C) \cap H \subseteq Z(H).$$

This yields (a). □

Lemma 1.14. Suppose that p is a prime, T is a p-subgroup of a group G, and M is a normal p'-subgroup of G. Let $C = C_G(T)$ and $N = N_G(T)$. Then

$$C_{G/M}(TM/M) = CM/M \text{ and } N_{G/M}(TM/M) = NM/M.$$

Proof. Let

$$C^*/M = C_{G/M}(TM/M) \text{ and } N^*/M = N_{G/M}(TM/M).$$

Clearly $NM \subseteq N^*$. On the other hand, take $x \in N^*$. Then x normalizes TM, so T^x is a Sylow p-subgroup of TM, and there exists $y \in M$ such that $T^x = T^y$. Then xy^{-1} normalizes T. Hence

$$xy^{-1} \in N \text{ and } x = (xy^{-1})y \in NM.$$

Thus $N^* = NM$. Now $CM \subseteq C^* \subseteq N^* = NM$. Since $T \cap M = 1$, we have $C^* \cap N = C$. Hence

$$C^* = (C^* \cap N)M = CM. \quad \square$$

Proposition 1.15. Suppose that G is a solvable group and p is a prime.

(a) **(P. Hall & G. Higman, "Lemma 1.2.3")** Assume that T is a Sylow p-subgroup of $\mathcal{O}_{p',p}(G)$. Then $C_G(T) \subseteq \mathcal{O}_{p',p}(G)$.

(b) **(D. Goldschmidt)** Assume that R is a p-subgroup of G. Then $\mathcal{O}_{p'}(C_G(R)) \subseteq \mathcal{O}_{p'}(G)$.

Proof. (a) **G**, Theorem 6.3.3, p. 228. (b) By Lemma 1.14, we may assume $\mathcal{O}_{p'}(G) = 1$. Let $M = \mathcal{O}_{p'}(C_G(R))$ and $T = \mathcal{O}_p(G)$. Then $RM = R \times M$ and M is an operator group on the p-group RT. Since $C_{RT}(R)$ normalizes M,

$$[C_{RT}(R), M] \subseteq RT \cap M = 1.$$

Therefore $C_{RT}(R)$ centralizes M. Let $C = C_{RT}(M)$. Then we have $C_{RT}(C) \subseteq C_{RT}(R) \subseteq C$. By Proposition 1.10, M centralizes T. As $T = F(G)$, we know $C_G(T) \subseteq T$. Thus $M = M \cap T = 1$. $\quad \square$

Proposition 1.16. Suppose that p is a prime, G is a p'-group, and A is a noncyclic abelian p-group of automorphisms of G. Then

$$G = \langle C_G(x) \mid x \in A^{\#} \rangle \text{ and}$$
$$G = \langle C_G(Y) \mid Y \subseteq A \text{ and } A/Y \text{ is cyclic} \rangle.$$

Proof. The first assertion is **G**, Theorem 6.2.4, p. 225. The second then follows by induction on $|G|$. $\quad \square$

Theorem 1.17 (D. G. Higman, "Focal Subgroup Theorem"). Suppose that G is a group, p is a prime, and S is a Sylow p-subgroup of G. Then

$$S \cap G' = \langle x^{-1}y \mid x, y \in S \text{ and } x \text{ is conjugate to } y \text{ in } G \rangle.$$

Proof. **G**, Theorem 7.3.4, p. 250. $\quad \square$

Theorem 1.18 (Burnside). Suppose that G is a group, p is a prime, and S is a Sylow p-subgroup of G. Assume that $S \subseteq Z(N_G(S))$. Then G has a normal p-complement.

Proof. Here S is abelian. Suppose x, $y \in S$ and $x^u = y$ for some $u \in G$. Then

$$S \subseteq C_G(x), \; S \subseteq C_G(y), \text{ and } S^u \subseteq C_G(x)^u = C_G(y).$$

By Sylow's Theorem, there exists $v \in C_G(y)$ such that $(S^u)^v = S$. Then

$$uv \in N_G(S) \text{ and } x^{uv} = (x^u)^v = y^v = y.$$

Since $S \subseteq Z(N_G(S))$, we know $x^{uv} = x$. Thus $x = y$ and $x^{-1}y = 1$.

By Theorem 1.17 and the argument above $S \cap G' = 1$, and hence G' is a p'-subgroup of G. Define a normal subgroup K of G by

$$K \supseteq G' \text{ and } K/G' = \mathcal{O}_{p'}(G/G').$$

It is easy to see that $KS = G$ and $K \cap S = 1$. Thus K is a normal p-complement in G. $\quad\square$

Corollary 1.19. Suppose that G is a group.

(a) If S is a cyclic Sylow subgroup of G, then either $S \cap G' = 1$ or $S \subseteq G'$.

(b) If G is a Z-group, then G' is a Hall subgroup of G.

Proof. Since (b) follows from (a), we will prove only (a).

Let K be a complement to S in $N_G(S)$. By Proposition 1.6(d), we know that $S = C_S(K) \times [S, K]$. Since S is cyclic, either $S = [S, K] \subseteq G'$, or $S = C_S(K) \subseteq Z(N_G(S))$. In the latter case, $S \cap G' = 1$ by Theorem 1.18. $\quad\square$

Theorem 1.20 (Maschke). Suppose that G is represented by linear transformations on a vector space V over a field $\mathbf{F}$. Assume that the characteristic of $\mathbf{F}$ is zero or is a prime that does not divide $|G|$.

Then V is completely reducible under G.

Proof. **G**, Theorem 3.3.2, p. 66. $\quad\square$

For later reference we gather together some elementary properties of p-length.

Lemma 1.21. Suppose that G is a finite group. Then:

(a) if G has p-length one and H is a subgroup of G, then H has p-length one;

(b) if H is a normal p'-subgroup of G and G/H has p-length one, then G has p-length one;

(c) if H is a normal p-subgroup of G such that $\mathcal{O}_{p'}(G/H) = 1$ and if G/H has p-length one, then G has p-length one;

(d) G has p-length one if and only if the subgroup of G generated by all p-elements of G has a normal p-complement; and

(e) if H and N are normal subgroups of G such that $H \cap N = 1$ and G/H and G/N both have p-length one, then G has p-length one.

Proof. (a), (b), and (c) are easily verified from the definition. For (d), let $U \lhd G$ be generated by all p-elements of G. If U has a normal p-complement K, then $K \operatorname{char} U$, so $K \lhd G$ and $K \subseteq \mathcal{O}_{p'}(G)$. Clearly $U\mathcal{O}_{p'}(G) = \mathcal{O}_{p',p}(G)$ and $G = \mathcal{O}_{p',p,p'}(G)$, as desired. Conversely, if $G = \mathcal{O}_{p',p,p'}(G)$, then $U \subseteq \mathcal{O}_{p',p}(G)$. Thus $\mathcal{O}_{p'}(G) \cap U \lhd U$ and

$$U/(\mathcal{O}_{p'}(G) \cap U) = U\mathcal{O}_{p'}(G)/\mathcal{O}_{p'}(G) \subseteq \mathcal{O}_{p',p}(G)/\mathcal{O}_{p'}(G),$$

which is a p-group. Thus $\mathcal{O}_{p'}(G) \cap U$ is a normal p-complement in U.

For (e), suppose G/H and G/N have p-length one, and let U be generated by all elements of p-power order in G. Applying (d) to G/H and G/N, we can find subgroups A and B of G such that $H \subseteq A \lhd UH$ and such that $N \subseteq B \lhd UN$ and A/H and B/N are normal p-complements in UH/H and UN/N, respectively. Clearly $A \cap B \cap U$ contains all of the p'-elements of U. On the other hand, if $g \in A \cap B$ is a p-element, then $g \in H$ and $g \in N$. Thus $g = 1$ and we conclude that $A \cap B \cap U$ is a normal p-complement in U. Now (d) yields (e). $\square$

Lemma 1.22. Suppose that p is a prime, G is a p-group, and $N \lhd G$. Let $|N| = p^k$. Then, for every nonnegative integer r such that $r \leq k$, N contains a normal subgroup of G having order p^r.

Proof. We use induction on $|G|$. The result is trivial if $N = 1$ or $r = 0$. Hence we assume that $N \neq 1$ and $r \geq 1$. Thus $N \cap Z(G) \neq 1$.

Take a subgroup Z of order p in $N \cap Z(G)$. By induction, N/Z contains a normal subgroup L/Z of G/Z having order p^{r-1}. Then $|L| = p^r$, $L \subseteq N$, and $L \lhd G$. $\square$

2. General Results on Representations

In this section, we consider representations of groups by matrices of finite degree or by finite-dimensional linear transformations. Assume G is a group. If G acts faithfully on a vector space V over a field $\mathbf{F}$, we denote the *enveloping algebra* of G over $\mathbf{F}$ by $E(G)$ (as in $\mathbf{G}$, p. 82). This is the smallest $\mathbf{F}$-subalgebra of $\operatorname{Hom}_{\mathbf{F}}(V, V)$ that contains G. As usual, we embed $\mathbf{F}$ in $\operatorname{Hom}_{\mathbf{F}}(V, V)$ by identifying field elements with scalar multiplications.

By *module* we will always mean finite-dimensional right module.

Suppose H is a subgroup of G. If L is an $\mathbf{F}G$-module, we denote the restriction of L to H by $L|_H$ or L_H. If M is an $\mathbf{F}H$-module, we denote by M^G the $\mathbf{F}G$-module induced from M. We consider M^G to be the tensor product $M \otimes_{\mathbf{F}H} \mathbf{F}G$. If, in addition, $H \lhd G$ and $x \in G$, we denote by M^x the $\mathbf{F}H$-module with underlying $\mathbf{F}$-module M and H-action (temporarily denoted by $*$) defined by

$$m * h = m(xhx^{-1}),$$

for all $m \in M$ and $h \in H$. The module M^x is called a *conjugate* $\mathbf{F}H$-module and is clearly isomorphic to the $\mathbf{F}H$-submodule $M \otimes x$ of M^G.

Proposition 2.1. Suppose that G is a group, $\mathbf{F}$ is a field, and M is an irreducible $\mathbf{F}G$-module. Then:

    (a) M is absolutely irreducible if and only if $\mathrm{Hom}_{\mathbf{F}G}(M, M) = \mathbf{F}$;

    (b) if G is faithful on M and $\mathrm{Hom}_{\mathbf{F}G}(M, M) = \mathbf{F}$, then $\mathrm{Hom}_{\mathbf{F}}(M, M) = E(G)$; and

    (c) if $\mathbf{F}$ is a finite field and $\mathbf{K} = \mathrm{Hom}_{\mathbf{F}G}(M, M)$, then $\mathbf{K}$ is a finite field and we can regard M as an absolutely irreducible $\mathbf{K}G$-module.

Proof. (a) If $\mathbf{F}$ has characteristic zero or relatively prime to $|G|$, this is $\mathbf{G}$, Theorem 3.5.7, p. 80. The general case can be deduced in one direction from the final paragraph of the proof in $\mathbf{G}$ (where one assumes that $D \neq \mathbf{F}$) and in the other from the Jacobson Density Theorem ($\mathbf{G}$, Theorem 3.6.2, p. 86). For a nice proof of the general case, see [3, Theorem 29.13, p. 202].

    (b) follows from the Jacobson Density Theorem ($\mathbf{G}$, Theorem 3.6.2, p. 86) and the fact that $\mathrm{Hom}_{\mathbf{F}G}(M, M) = \mathrm{Hom}_{E(G)}(M, M)$.

    (c) By Schur's Lemma ($\mathbf{G}$, Theorem 3.5.2, p. 76), $\mathbf{K}$ is a division algebra with $\mathbf{F}$ in its center. Since $\mathbf{F}$ is finite and M has finite dimension, M is actually finite. Therefore $\mathbf{K}$ is also finite and, by Wedderburn's well-known theorem on finite division rings [16, Theorem 7.2.1, p. 361], $\mathbf{K}$ is a field. Since $\mathbf{K} = \mathrm{Hom}_{\mathbf{F}G}(M, M)$, we can regard M as a vector space over $\mathbf{K}$ and the elements of G as linear transformations of M over $\mathbf{K}$. Clearly

$$\mathbf{K} \subseteq \mathrm{Hom}_{\mathbf{K}G}(M, M) \subseteq \mathrm{Hom}_{\mathbf{F}G}(M, M) = \mathbf{K}.$$

By (a), M is an absolutely irreducible $\mathbf{K}G$-module. $\square$

Proposition 2.2. Suppose that G is a group, $H \lhd G$, and G/H is cyclic. Assume that $\mathbf{F}$ is an algebraically closed field and M is an irreducible $\mathbf{F}H$-module such that $M \cong M^x$ for all $x \in G$.

    (a) If L is an irreducible $\mathbf{F}G$-module and M is isomorphic to a submodule of L_H, then $L_H \cong M$.

    (b) The representation of H on M can be extended to a representation of G.

Proof. (a) We can assume that G acts faithfully on L. By Clifford's Theorem ($\mathbf{G}$, Theorem 3.4.1, p. 70), there exists an integer k such that

$$(2.1) \qquad\qquad L_H = M_1 \oplus M_2 \oplus \cdots \oplus M_k,$$

where $M \cong M_i$ for each i. Since G acts faithfully on L, H acts faithfully on M.

    Consider, for a moment, the action of H on M. Since $\mathbf{F}$ is algebraically closed, $\mathrm{Hom}_{\mathbf{F}H}(M, M) = \mathbf{F}$ and, by Proposition 2.1, $E(H) = \mathrm{Hom}_{\mathbf{F}}(M, M)$. Take $x \in G$ such that $G = \langle H, x \rangle$. Then, by hypothesis, $M \cong M^{x^{-1}}$, and therefore there is an $\mathbf{F}$-isomorphism $\tau \in \mathrm{Hom}_{\mathbf{F}}(M, M) = E(H)$ such that for all $m \in M$ and $h \in H$,

$$(2.2) \qquad\qquad (mh)\tau = (m\tau)xhx^{-1}.$$

On the other hand, $H \subseteq G$, and H acts on the module L. By (2.1) and (2.2), there exists an $\mathbf{F}$-isomorphism (which we also call τ) such that $\tau \in \mathrm{E}(H) \subseteq \mathrm{E}(G) \subseteq \mathrm{Hom}_{\mathbf{F}}(L, L)$ and, for all $\ell \in L$ and $h \in H$,

$$(\ell h)\tau = (\ell\tau)xhx^{-1}.$$

Extending linearly, we see that, for any $\ell \in L$ and any $\theta \in \mathrm{E}(H) \subseteq \mathrm{E}(G)$,

$$(\ell\theta)\tau = (\ell\tau)x\theta x^{-1}.$$

Thus

(2.3) $(\ell\theta)\tau x = (\ell\tau x)\theta.$

In particular, since $\tau^{-1} \in \mathrm{E}(H)$, we have

$$\ell x = \ell\tau^{-1}\tau x = \ell\tau x\tau^{-1}$$

and therefore

(2.4) $(\ell x)\tau x = (\ell\tau x\tau^{-1})\tau x = (\ell\tau x)x.$

Since H and x generate G, (2.3) and (2.4) imply that $\tau x \in \mathrm{Hom}_{\mathbf{F}G}(L, L)$. Since $\mathbf{F}$ is algebraically closed and L is irreducible, Proposition 2.1 implies that $\mathrm{Hom}_{\mathbf{F}G}(L, L) = \mathbf{F}$. Thus τx is a scalar. Since $\tau \in \mathrm{E}(H)$, we have $M_1\tau = M_1$. Hence

$$M_1 = M_1\tau x = M_1 x.$$

Thus M_1 is a G-submodule of L and consequently $L = M_1$.

(b) Let L be an irreducible $\mathbf{F}G$-submodule of M^G. Then L_H is a direct sum of copies of M. By (a), $L_H \cong M$. □

Lemma 2.3. Suppose that G is a solvable group, $\mathbf{F}$ is a field, and M is an absolutely irreducible $\mathbf{F}G$-module. Then $\dim M$ divides $|G|$.

Remark. This lemma is a corollary of a well-known theorem of Fong and Swan [5, Theorem 72.1, p. 473].

Proof. Use induction on $|G|$. We may assume that $|G| > 1$ and $\mathbf{F}$ is algebraically closed. Take $H \lhd G$ of some prime index p. Let L be an irreducible submodule of M_H. By induction,

(2.5) $\dim L$ divides $|H|$.

Take $x \in G - H$. If $L \cong L^x$, then, by Proposition 2.2, $L = M_H$, and we are done. Otherwise

(2.6) $M_H = L + Lx + \cdots + Lx^{p-1},$

and the $\mathbf{F}H$-submodules $L, Lx, \ldots, Lx^{p-1}$ are pairwise nonisomorphic. In this case the sum in (2.6) is direct and $\dim M = p(\dim L)$. Then we are done by (2.5). □

The next proposition uses only elementary techniques of linear algebra.

Proposition 2.4. Suppose that V is a vector space over a field $\mathbf{F}$ and $\dim V = q \geq 2$. Let g be an invertible linear transformation of V of finite order $h \geq 2$ and assume that $\mathbf{F}$ contains a primitive h^{th} root of unity ϵ. For all integers i and t define

$$E = \text{End}_{\mathbf{F}}(V),$$
$$V_i = \left\{ v \in V \mid vg = \epsilon^i v \right\},$$
$$n_i = \dim V_i,$$
$$E_i = \left\{ e \in E \mid e^g = g^{-1}eg = \epsilon^i e \right\}, \text{ and}$$
$$E_{i,t} = \left\{ e \in E \mid V_i e \subseteq V_t \text{ and } V_j e = 0 \text{ for all } j \neq i \right\}.$$

Then:
- (a) $V = V_0 \oplus V_1 \oplus \cdots \oplus V_{h-1}$;
- (b) $n_i = n_{i+h}$ for all i;
- (c) $E = \displaystyle\bigoplus_{0 \leq i,t \leq h-1} E_{i,t}$
- (d) $\dim E_{i,t} = n_i n_t$ for all i and t;
- (e) $E_{i,t} \subseteq E_{t-i}$ for all i and t;
- (f) $E_m = \displaystyle\bigoplus_{\substack{t-i \equiv m \ (\text{mod } h) \\ 0 \leq i,t \leq h-1}} E_{i,t}$ for all m;
- (g) $\dim E_m = \displaystyle\sum_{i=0}^{h-1} n_i n_{i+m}$ for all m;
- (h) $2 \dim E_0 - 2 \dim E_m = \displaystyle\sum_{i=0}^{h-1} (n_i - n_{i+m})^2$ for all m;
- (j) if $\dim E_0 = \dim E_m + 1$ for all $m \not\equiv 0 \ (\text{mod } h)$, then there exist integers i, n, and $\delta = \pm 1$ such that $q = hn + \delta$, $n_i = n + \delta$, and $n_j = n$ for all $j \not\equiv i \ (\text{mod } h)$; and
- (k) under the same assumptions as (j), $\dim V_0 = n_0 > 0$ unless $n = 1$, $i = 0$, $\delta = -1$, and $h = q + 1$.

Proof. The assumption that $\mathbf{F}$ contains a primitive h^{th} root of unity forces $\mathbf{F}$ to have characteristic not dividing h.

Statements (a) and (b) are clear. For (c) and (d), it may help to consider the matrices of elements of $\text{Hom}_{\mathbf{F}}(V, V)$ with respect to a basis that is the union of bases of the subspaces V_i. Each such matrix A can be viewed naturally as a matrix of $h \times h$ submatrices A_{it} with n_i rows and n_t columns.

For (e), take $e \in E_{i,t}$ and $v \in V_i$. Clearly $e^g = g^{-1}eg \in E_{i,t}$ and $ve^g = vg^{-1}eg = (\epsilon^{-i}v)eg = \epsilon^{-i}(ve)g = \epsilon^{-i}\epsilon^t(ve) = v(\epsilon^{t-i}e)$. Thus $e^g = \epsilon^{t-i}e$.

Now (f) follows from (c) and (e); (g) from (d) and (f); and (h) from (b) and (g).

To prove (j), assume that $\dim E_0 = \dim E_m + 1$ for every $m \not\equiv 0 \pmod{h}$. Let

$$S_1 = \{\, i \mid 0 \le i \le h - 1 \text{ and } n_i = n_0 \,\} \text{ and}$$
$$S_2 = \{\, i \mid 0 \le i \le h - 1 \text{ and } n_i \ne n_0 \,\}.$$

By (h) with $m = 1$, S_2 is not empty. Let j and k be the smallest and largest elements of S_2, respectively. (They may be equal.) Then $n_i = n_0$ for $i = 1, 2, \ldots, j - 1$ and for $i = k + 1, \ldots, h - 1$. By (b), $n_h = n_0$. Therefore, by (h) with $m = 1$,

$$2 = (n_{j-1} - n_j)^2 + (n_j - n_{j+1})^2 + \cdots + (n_k - n_{k+1})^2,$$

which yields: $|n_0 - n_j| = |n_k - n_0| = 1$ and $n_j = n_{j+1} = \cdots = n_k$. Consequently, $|S_2| = k + 1 - j$ and $|S_1| = h - (k + 1 - j) = h + j - 1 - k$.

If $|S_1| = 1$, we may take $i = 0$, $n = n_1$, and $\delta = n_0 - n$; and similarly if $|S_2| = 1$. But, suppose $|S_1|, |S_2| \ge 2$. Then $h \ge 4$ and $j + 1 \le k \le h + j - 3$. Hence, by (b), $n_{k+1} = n_{k+2} = n_0$. Therefore, by (h) with $m = 2$,

$$2 \ge (n_{j-1} - n_{j+1})^2 + (n_{k-1} - n_{k+1})^2 + (n_k - n_{k+2})^2 = 3(n_0 - n_j)^2 = 3,$$

a contradiction.

Finally, under the assumptions for (j), we have $n \ge 1$, because $hn + \delta = q \ge 2$. Thus (k) follows from (j). $\square$

Remark. We thank Curtis Bennett for suggesting a simplification of our original proof of Proposition 2.4.

Theorem 2.5. Suppose that P is an extraspecial p-group of order p^{2n+1} for some prime p. Let G be the semidirect product of P (with $P \lhd G$) and a cyclic group H of order $|H| = h$ such that h is relatively prime to p and for all $x \in H^{\#}$,

$$C_P(x) = Z(P).$$

Suppose $\mathbf{F}$ is a field whose characteristic does not divide $|G|$. Then h divides $p^n + 1$ or $p^n - 1$ and, if $h \ne p^n + 1$, then every faithful, irreducible $\mathbf{F}G$-module V satisfies

$$C_V(H) \ne 0.$$

Remark. The last part of the theorem fails if one allows $h = p^n + 1$. There are counterexamples in which $p^n = 2$, $h = 3$, $G \cong \mathbf{SL}(2, 3)$, and $\mathbf{F}$ is an arbitrary field of prime characteristic not dividing $|G|$, or an algebraically closed field of characteristic zero.

Proof. Since the characteristic of $\mathbf{F}$ does not divide $|G|$, it is easy to see that there exists an irreducible $\mathbf{F}G$-module on which $Z(P)$ acts nontrivially. Suppose V is any such module (e.g., a faithful $\mathbf{F}G$-module). Since $C_V(Z(P))$ is invariant under G,

$$(2.7) \qquad\qquad C_V(Z(P)) = 0.$$

Let $\mathbf{F}^*$ be the algebraic closure of $\mathbf{F}$ and $V^* = \mathbf{F}^* \otimes_{\mathbf{F}} V$. Take a generator x of H. Then

$$C_V(H) = C_V(x) \text{ and}$$
$$\dim C_V(x) = \dim V - \operatorname{rank}(x-1),$$

where $x-1$ denotes the linear transformation of V given by $v(x-1) = vx - v$ for all $v \in V$. Since extending $\mathbf{F}$ to $\mathbf{F}^*$ does not change the rank of $x-1$, we have

$$(2.8) \qquad\qquad \dim_{\mathbf{F}^*} C_{V^*}(H) = \dim_{\mathbf{F}} C_V(H).$$

A similar argument, together with (2.7), shows that

$$(2.9) \qquad\qquad C_{V^*}(Z(P)) = 0.$$

Let W be an irreducible G-submodule of V^* and M an irreducible P-submodule of V^*. Since P is extraspecial, $|Z(P)| = p$. Therefore every nonidentity normal subgroup of P contains $Z(P)$. Consequently (2.9) implies that

$$(2.10) \qquad\qquad P \text{ acts faithfully on } M \text{ and on } W.$$

This shows that $[C_G(W), P] \subseteq C_G(W) \cap P = 1$. Hence

$$C_G(W) = C_G(W) \cap C_G(P) = C_G(W) \cap Z(P) = 1.$$

Thus G acts faithfully and irreducibly on W. Clearly it suffices to prove that h divides $p^n \pm 1$ and that if $h \neq p^n + 1$, then $C_W(H) \neq 0$, for, by (2.8),

$$\dim_{\mathbf{F}} C_V(H) = \dim_{\mathbf{F}^*} C_{V^*}(H) \geq \dim_{\mathbf{F}^*} C_W(H).$$

Thus we may assume that $\mathbf{F} = \mathbf{F}^*$ and $W = V$. Then M is an irreducible P-submodule of V.

By (2.10), P is faithful on M. Furthermore, by **G**, Theorem 5.5.4, p. 206 and the discussion following the theorem, a faithful, irreducible representation of an extraspecial group is determined by the action of its center. It follows that, for any $g \in G$, the P-submodules M and M^g are isomorphic. By Proposition 2.2(a), $V_P = M$ and, by **G**, Theorem 5.5.5, p. 208, $\dim M = \dim V = p^n$. Let $q = p^n$.

By Proposition 2.1, $E(P) = \mathrm{Hom}_{\mathbf{F}}(V, V)$. Identifying elements of P with their images in $E(P)$, we obtain

$$E(P) = \sum_{g \in P} \mathbf{F}g$$

and, since elements of $Z(P)$ act as scalars,

(2.11)
$$E(P) = \sum_{g \in R} \mathbf{F}g,$$

where R is a set of coset representatives of $Z(P)$ in P. Taking $E = E(P)$ and $Z = Z(P)$, we have $|R| = |P/Z| = p^{2n} = q^2 = \dim(E)$. Thus the sum (2.11) is direct, and R is a basis of E.

By Proposition 1.5, $C_{P/Z}(x) = C_P(x)Z/Z = 1$ for every $x \in H^{\#}$. Hence, for each element $a \in P - Z$, we know that a and all of its H-conjugates lie in different cosets of Z. Thus we can fix a set of coset representatives R consisting of the element 1 and $(q^2 - 1)/h$ H-classes of P.

If we consider E as an H-module under conjugation, the direct decomposition (2.11) shows that E is the direct sum of the principal module and $(q^2 - 1)/h$ copies of the regular module. Since H is abelian, each irreducible H-module occurs with multiplicity one in the regular module, and hence the multiplicity of the principal module in E is one more than the multiplicity of any nontrivial irreducible module.

Taking g to be any generator of H, we see that the hypothesis of Proposition 2.4(j) is satisfied. Therefore, for some $\delta = \pm 1$ and some integer n', $p^n = q = hn' + \delta$, so that h divides $p^n - \delta$. Moreover, by Proposition 2.4(k), if $C_V(H) = 0$, then $h = q + 1 = p^n + 1$. $\square$

Theorem 2.6. Suppose that G is a finite group of odd order, $\mathbf{F}$ is a field, and V is an $\mathbf{F}G$-module of dimension two over $\mathbf{F}$ on which G acts faithfully. Then:

  (a) if the characteristic of $\mathbf{F}$ does not divide $|G|$, then G is abelian; and
  (b) if the characteristic of $\mathbf{F}$ is a prime divisor p of $|G|$, then G has an abelian Sylow p-subgroup that contains G'.

Proof. We use induction on $|G|$. We may regard G as a subgroup of $\mathrm{GL}(V, \mathbf{F})$. Let $G^* = G \cap \mathrm{SL}(V, \mathbf{F})$. By considering the tensor product of V with the algebraic closure of $\mathbf{F}$, we may assume that $\mathbf{F}$ is algebraically closed. Let p be the characteristic of $\mathbf{F}$.

Suppose $\mathcal{O}_q(G^*) \neq 1$ for some prime q. Let $K = \Omega_1(Z(\mathcal{O}_q(G^*)))$. Then K is an elementary abelian q-group and $K \triangleleft G$. We consider separately the cases in which $q = p$ and $q \neq p$.

Suppose $q = p$. Let $W = C_V(K)$. By $\mathbf{G}$, Lemma 2.6.3, p. 31, $W \neq 0$. Since V is 2-dimensional and G is faithful on V,

(2.12)
$$\dim W = \dim V/W = 1.$$

Now W is invariant under G. Let $C = C_G(W) \cap C_G(V/W)$. Simple matrix calculations show that C is an elementary abelian p-group. Since $\mathbf{F}$ has characteristic p, the only element of the multiplicative group $\mathbf{F} - \{0\}$ of p-power order is 1. Thus C contains every p-element of G. Similarly, C contains G' because $\mathbf{F} - \{0\}$ is abelian. Hence in this case we have (b).

Now suppose $q \neq p$. Then V is a direct sum of irreducible $\mathbf{F}K$-modules (by Theorem 1.20). Since K is abelian and $\mathbf{F}$ is algebraically closed, by **G**, Theorem 3.2.4, p. 65, all irreducible $\mathbf{F}K$-modules are one-dimensional. Thus $V = W_1 \oplus W_2$ for two one-dimensional $\mathbf{F}K$-modules W_1 and W_2. Take $x \in K^{\#}$, $w_1 \in W_1^{\#}$, and $w_2 \in W_2^{\#}$. Then

$$w_1 x = \lambda_1 w_1 \text{ and } w_2 x = \lambda_2 w_2 \text{ for some } \lambda_1, \lambda_2 \in \mathbf{F}.$$

Moreover, $\lambda_1 \lambda_2 = \det x = 1$, because $x \in G^* = G \cap \mathbf{SL}(V, \mathbf{F})$. Since x has odd order, $\lambda_1 \neq \lambda_2$. Now an easy calculation (or elementary result on modules) shows that W_1 and W_2 are the *only* one-dimensional subspaces of V fixed by x and thus the only one-dimensional $\mathbf{F}K$-submodules of V. Since $K \lhd G$, every element of G fixes or interchanges W_1 and W_2. As $|G|$ is odd, every element of G fixes W_1 and W_2. Therefore G is an abelian p'-group. Thus (a) applies here.

Now assume more generally that $G^* \neq 1$. If G^* is a p-group, then $\mathcal{O}_p(G^*) = G^* \neq 1$. On the other hand, assume that $|G^*|$ is divisible by some prime q different from p. Let Q be a Sylow q-subgroup of G^* and $H = N_{G^*}(Q)$. Then $\mathcal{O}_q(H) \neq 1$. The previous paragraph shows that H is an abelian group, so that Q is in the center of its normalizer in G^*. By a theorem of Burnside (Theorem 1.18), G^* has a normal complement N to Q. If $N = 1$, then $G^* = Q$ and $1 \subset \mathcal{O}_q(G^*) = Q$. If $N \neq 1$, then by induction, $\mathcal{O}_r(N) \neq 1$ for some prime r, and then $1 \subset \mathcal{O}_r(N) \subseteq \mathcal{O}_r(G^*)$. Either way, by the previous arguments, we are done.

Finally, assume that $G^* = 1$. Since $\mathbf{GL}(V, \mathbf{F})/\mathbf{SL}(V, \mathbf{F}) \cong \mathbf{F} - \{0\}$ under multiplication and $G^* = G \cap \mathbf{SL}(V, \mathbf{F})$, we see that G is an abelian p'-group. $\square$

Lemma 2.7. Suppose that p and q are distinct primes, P and Q are elementary abelian groups of order p^2 and q^2 respectively, and $Q \subseteq \mathrm{Aut}(P)$. Then

   (a) q divides $(p-1)$, and
   (b) there exists an element $\alpha \in Q^{\#}$ and an integer r such that

$$x^{\alpha} = x^r \text{ for every } x \in P, \ r^q \equiv 1 \pmod{p}, \text{ and } r \not\equiv 1 \pmod{p}.$$

Proof. We may regard P as a 2-dimensional vector space over $\mathbf{F}_p$ and Q as a group of linear transformations of P over $\mathbf{F}_p$. Since Q is abelian but not cyclic, Q is not irreducible on P (by **G**, Theorem 3.2.3). Therefore P is the direct sum of two 1-dimensional Q-submodules P_1 and P_2.

Take $0 \neq v_i \in P_i$, $i = 1, 2$. Now an easy argument shows that Q is the set of all linear transformations β of P with the property that

$$v_1^\beta = \lambda_1 v_1 \text{ and } v_2^\beta = \lambda_2 v_2 \text{ for some } \lambda_1, \lambda_2 \in \mathbf{F}_p \text{ such that } \lambda_1^q = \lambda_2^q = 1,$$

and then (a) and (b) follow. $\square$

3. Actions of Frobenius Groups and Related Results

Lemma 3.1. Suppose that K and R are nonidentity subgroups of a group G such that $K \lhd G$, $KR = G$, and $K \cap R = 1$. Then the following are equivalent:

(a) G is a Frobenius group with Frobenius complement R and Frobenius kernel K;

(b) $C_K(x) = 1$ for all $x \in R^\#$.

Proof. By **G**, Theorem 2.7.6, p. 38, (a) yields (b). Now assume (b). Take any $y \in G$ such that $R \cap R^y \neq 1$. Let $h \in (R \cap R^y)^\#$. Choose $u \in R$ and $v \in K$ such that $uv = y$. Then

$$h \in R \cap R^y = R \cap R^{uv} = R \cap R^v.$$

Thus $h = h_0^{\,v} = v^{-1} h_0 v$ for some $h_0 \in R$. Since $K \lhd G$, it follows that $h_0 v h_0^{-1} \in K$, and therefore

$$h h_0^{-1} = v^{-1} h_0 v h_0^{-1} \in R \cap K = 1.$$

Hence $h = h_0$. Now $v \in C_K(h)$ so, by (b), $v = 1$ and $y = uv = u \in R$. Thus we have shown that

$$R \cap R^y = 1 \text{ whenever } y \in G - R.$$

Therefore $N_G(R) = R$, and R is disjoint from its distinct conjugates in G. By **G**, Theorem 2.7.7, p. 39, G is a Frobenius group with Frobenius complement R. By hypothesis,

$$K \cap R^g = K^g \cap R^g = (K \cap R)^g = 1 \text{ for all } g \in G.$$

Thus K is contained in the Frobenius kernel of G, which has order $|G : R|$. Since $|K| = |G : R|$, K is the Frobenius kernel of G. $\square$

Lemma 3.2. Let $G = KR$ be a Frobenius group with solvable Frobenius kernel K and Frobenius complement R and suppose that $N \lhd G$. Assume that $K \not\subseteq N$. For each subgroup H of G let $\overline{H} = HN/N$. Then:

(a) $N \subset K$; and

(b) $\overline{G}$ is a Frobenius group with Frobenius kernel $\overline{K}$ and Frobenius complement $\overline{R}$.

Note. Since Thompson's Thesis (**G**, Theorem 10.2.1, p. 337) implies that the kernel of a Frobenius group is nilpotent (**G**, Theorem 10.3.1(iii), p. 339), the assumption that K is solvable is unnecessary.

Proof. Since G is a Frobenius group with kernel K and complement R, the conjugates R^x for $x \in K$ are pairwise disjoint (except for the identity) and

$$(3.1) \qquad\qquad G = K \cup \bigcup_{x \in K} R^x.$$

First suppose that $N \subseteq K$. Since $K \not\subseteq N$, we know that $N \subset K$, which is (a). Since G is a Frobenius group, by Lemma 3.1, $C_K(x) = 1$ for every $x \in R^\#$. By Proposition 1.5(d),

$$C_{\overline{K}}(x) = C_K(x)N/N = 1,$$

and a second application of Lemma 3.1 implies that $\overline{G}$ is a Frobenius group with kernel $\overline{K}$ and complement $\overline{R}$, as desired.

Now let N be an arbitrary normal subgroup of G such that $K \not\subseteq N$. By the previous paragraph, it suffices to show that $N \subseteq K$. Set $H = N \cap K$, $\widehat{G} = G/H$ and, for every subgroup $L \subseteq G$, let $\widehat{L} = LH/H$. By the argument above, $\widehat{G}$ is a Frobenius group with kernel $\widehat{K}$ and complement $\widehat{R}$. By **G**, Theorem 2.7.7, p. 39, $\widehat{R}^x \cap \widehat{R} = 1$ for every $x \in \widehat{K}^\#$. Thus $(\widehat{N \cap R})^x \cap (\widehat{N \cap R}) = 1$ for every $x \in \widehat{K}^\#$.

On the other hand, $[N \cap R, K] \subseteq N \cap K = H$ and so $[\widehat{N \cap R}, \widehat{K}] = 1$. Therefore $(\widehat{N \cap R})^x \subseteq (\widehat{N \cap R})$ for all $x \in \widehat{K}$. Since $\widehat{K} \neq 1$, this implies that $\widehat{N \cap R} = 1$. Thus $(N \cap R) \subseteq H \subseteq K$ and consequently $N \cap R = 1$. But $N \lhd G$, so, for every $x \in K$,

$$N \cap R^x = (N \cap R)^x = 1.$$

Finally (3.1) implies that $N \subseteq K$. $\square$

Lemma 3.3. Let $G = KR$ be a Frobenius group with Frobenius kernel K and Frobenius complement R. Suppose that G is represented on a vector space V over a field $\mathbf{F}$ of characteristic not dividing $|K|$. If K does not act trivially on V, then $C_V(R) \neq 0$.

Proof (Wielandt). The representation of G on V induces a unique representation of the group algebra $\mathbf{F}G$ on V. For any subgroup H of G, consider the element $\overline{H} = \sum_{h \in H} h \in \mathbf{F}G$. Clearly, for any $v \in V$, we have $v\overline{H} \in C_V(H)$.

Suppose that $C_V(R) = 0$. We will show that the action of K on V is trivial. For any $g \in G$ we have $C_V(R^g) = 0$ and $C_V(G) \subseteq C_V(R) = 0$. Thus, for all $v \in V$, $v\overline{R} = v\overline{R^g} = v\overline{G} = 0$.

Since G is a Frobenius group with kernel K and complement R, the conjugates R^x for $x \in K$ are pairwise disjoint (except for the identity) and

$$G = K \cup \bigcup_{x \in K} R^x$$

3. Actions of Frobenius Groups and Related Results

where, except for the identity, the union is disjoint. Thus

$$\overline{G} = \overline{K} + \sum_{x \in K} \overline{R^x} - |K|1$$

and hence, for any $v \in V$,

$$0 = v\overline{K} - |K|v.$$

But then $|K|v = v\overline{K} \in C_V(K)$ for all $v \in V$. Since $|K| \neq 0$ in $\mathbf{F}$, this implies that $V \subseteq C_V(K)$, and hence K acts trivially on V. $\square$

Theorem 3.4. Let G be a solvable group of odd order that has a normal Hall subgroup K with a complement R of prime order. Suppose that G acts on a vector space V over a field $\mathbf{F}$ whose characteristic does not divide $|G|$. If $C_V(R) = 0$, then $[R, K] \subseteq C_K(V)$.

Remark. The counterexamples for $\mathbf{F}$ finite mentioned in the remark following Theorem 2.5 yield counterexamples here if one allows G to have even order. In particular, for every prime p greater than three, there is a counterexample in which $G \cong \mathbf{SL}(2,3)$ and $\mathbf{F} = \mathbf{F}_p$.

Proof. Suppose false and let G be a minimal counterexample. Let $C = C_G(V)$. If H is any proper R-invariant subgroup of K, the group HR satisfies the hypotheses of the theorem and $|HR| < |G|$. By minimality of G, we have $[H, R] \subseteq C$. Therefore, since, by assumption, $[R, K] \not\subseteq C$,

(3.2) K is not generated by its proper R-invariant subgroups.

Since the characteristic of $\mathbf{F}$ does not divide $|G|$, by Maschke's Theorem (Theorem 1.20), we know that V is the direct sum of irreducible submodules. Then, since $[K, R] \not\subseteq C$, we can choose an irreducible G-submodule M such that $[K, R]$ does not act trivially on M. Let $N = C_G(M)$ and $p = |R|$. Since we know that $C_M(R) \subseteq C_V(R) = 0$, we obtain $N \cap R = 1$. Thus $|NR| = |N||R| = p|N|$ and $|G| = p|K|$. Hence N is a normal p'-subgroup of G and $N \subset K = \mathcal{O}_{p'}(G)$.

Let $\overline{G} = G/N$, $\overline{K} = K/N$, and $\overline{R} = RN/N$. Then $\overline{K}$ is a normal Hall subgroup of $\overline{G}$ with complement $\overline{R}$ and $C_M(\overline{R}) = 0$. Furthermore, $\overline{G}$ acts faithfully on M. If $N \neq 1$, then $|\overline{G}| < |G|$ and, by minimality of G, we conclude that $[\overline{K}, \overline{R}] = 1$. But then $[K, R] \subseteq N$, which contradicts the choice of M. Thus $N = 1$, and we conclude that G acts faithfully and irreducibly on M, and hence faithfully on V. By **G**, Theorem 3.2.2, p. 64,

$$Z(G) \text{ is cyclic.}$$

Suppose that $|K|$ is divisible by two or more distinct primes. By Proposition 1.5(a), K has R-invariant Sylow p-subgroups for every prime p dividing $|K|$. But then K is generated by proper R-invariant subgroups, which contradicts (3.2). Thus $|K| = q^n$ for some integer n and prime q.

By **G**, Theorem 5.3.7, p. 181, K is either an elementary abelian group or a nonabelian special group, and

(3.3) R acts irreducibly on K/K' and centralizes K'.

Suppose K is elementary abelian. Then $K' = 1$. Furthermore, since $[R, K] \neq 1$, we have $C_K(R) \subset K$. Therefore, by (3.3), $C_K(R) = 1$. But now, by Lemma 3.1, G is a Frobenius group with kernel K and complement R. Since G acts faithfully on V, the action of K is nontrivial and Lemma 3.3 implies that $C_V(R) \neq 0$, which contradicts our original assumption. Thus K is a nonabelian special group.

Now, since K is a nonabelian special group, $Z(K) = K'$, and therefore, by (3.3), R centralizes $Z(K)$. Thus $Z(K) \subseteq Z(G)$, and hence $Z(K)$ is cyclic. Consequently K is extraspecial.

By (3.3), R acts irreducibly on $K/K' = K/Z(K)$ and $Z(K)$ is a maximal R-invariant subgroup of K. Thus $Z(K) = C_K(R) = C_K(x)$ for all $x \in R^{\#}$. We can now apply Theorem 2.5 to obtain a final contradiction.   □

Theorem 3.5. Let $G = KR$ be a Frobenius group with solvable Frobenius kernel K and cyclic Frobenius complement R of prime order. Suppose that G acts on a vector space V over a field $\mathbf{F}$ whose characteristic does not divide $|G|$. If $C_V(R)$ is one-dimensional, then $K' \subseteq C_K(V)$.

Note. Just as for Lemma 3.2, Thompson's Thesis implies that K is actually nilpotent, so the assumption that K is solvable is not necessary.

Proof. Proceed by induction on $|G|$. Suppose G satisfies the hypotheses of the theorem. The hypotheses remain intact if we replace $\mathbf{F}$ by its algebraic closure $\overline{\mathbf{F}}$ and V by $V \otimes_{\mathbf{F}} \overline{\mathbf{F}}$, so, without loss of generality, we can assume that $\mathbf{F}$ is algebraically closed.

Suppose that G does not act faithfully on V. Let $C = C_G(V)$. Then $1 \subset C \lhd G$. If $K \subseteq C$, then $K' \subseteq C$, as desired. Assume then that $K \not\subseteq C$. By Lemma 3.2, $C \subset K$ and G/C is a Frobenius group with Frobenius complement RC/C and Frobenius kernel K/C. Since $C_V(RC/C) = C_V(R)$, induction yields

$$K'C/C = (K/C)' \subseteq C_{G/C}(V) = C/C.$$

Thus $K' \subseteq C$, as desired.

For the rest of the proof assume that G acts faithfully on V.

Suppose N is a proper R-invariant subgroup of K. Then NR satisfies the hypotheses of the theorem and $|NR| < |G|$. By induction, N is abelian. Hence

(3.4)   every proper R-invariant subgroup of K is abelian.

Since K is solvable, K' is a proper characteristic subgroup of K. By (3.4), K' is abelian.

Suppose $N \subseteq K$ is an R-invariant subgroup of K and $U \oplus W \subseteq V$ is an NR-submodule of V with U and W both NR-invariant. If N acts nontrivially on U, then, by Lemma 3.3, $C_U(R) \neq 0$. But then, since $C_V(R)$ is one-dimensional, $C_V(R) = C_U(R) \subseteq U$. On the other hand, $C_W(R) = C_V(R) \cap W \subseteq U \cap W = 0$, so a second application of Lemma 3.3 implies that N acts trivially on W. Thus, for U, W, and N as in the first sentence of this paragraph,

$$(3.5) \qquad \text{either } U \subseteq C_V(N) \text{ or } W \subseteq C_V(N).$$

Since G acts faithfully on V, there exists an irreducible G-submodule U of V on which K acts nontrivially. Since the characteristic of $\mathbf{F}$ does not divide $|G|$, by Maschke's Theorem (Theorem 1.20), V is completely reducible and U has a G-invariant complement W in V. By (3.5), we have $W \subseteq C_V(K) \subseteq C_V(K')$.

Suppose that $C_U(K') \neq 0$. Since $K' \lhd G$, $C_U(K')$ is a submodule of U. Since U is irreducible, this implies $U = C_U(K') \subseteq C_V(K')$. But then $V = U \oplus W \subseteq C_V(K')$ and, since G is faithful on V, we have $K' = 1$, as desired. Thus it suffices to prove that $C_U(K') \neq 0$.

Consider the structure of U as a K-module. Let M be an irreducible K-submodule of U and suppose $R = \langle x \rangle$ with $|R| = p$. Now Clifford's Theorem ($\mathbf{G}$, Theorem 3.4.1, p. 70) implies that U is the direct sum of conjugates of M and R permutes the Wedderburn components of U transitively. Thus either U has only one Wedderburn component (in which case the conjugates of M are all isomorphic) or U has p Wedderburn components (in which case the conjugates of M are all distinct). In the latter case, we can write $U = M \oplus M^x \oplus \cdots \oplus M^{x^{p-1}}$. Let $\pi_1 \colon U \to M$ be projection onto M. For any $m \in M$, the element $m + mx + \cdots + mx^{p-1}$ is fixed by R. Thus π_1 takes $C_U(R)$ onto M. But $C_U(R)$ is one-dimensional, so M is a one-dimensional K-module. Therefore $0 \subset M \subseteq C_U(K')$ and consequently $C_U(K') \neq 0$. Thus we can assume the conjugates of M are all isomorphic. In this case, Proposition 2.2 implies that $U_K \cong M$, so U is an irreducible K-module.

Now consider the structure of U as a $K'R$-module. If U is reducible, (3.5) implies $C_U(K') \neq 0$, so we can assume that U is irreducible as a $K'R$-module.

Finally, we consider the structure of U as a K'-module. By Clifford's Theorem, U is the direct sum of Wedderburn components with respect to K'. On the one hand, R permutes these components transitively, so the number of Wedderburn components divides p. On the other hand, K permutes these Wedderburn components transitively and the number of components divides $|K|$. Since p is relatively prime to $|K|$, this implies that there is only one Wedderburn component. A second application of Proposition 2.2 implies that U is an irreducible K'-module.

Since $\mathbf{F}$ is algebraically closed and K' is abelian, U is one-dimensional. Consequently $U \subseteq C_U(K')$, and $C_U(K') \neq 0$, as desired. $\square$

Recall that a group G is called a Z-group if all of its Sylow subgroups are cyclic.

Theorem 3.6. Let G be a solvable group of odd order and suppose that H is a normal Hall subgroup of G. Let R be a complement of H in G and suppose that R_0 is a subgroup of R of prime order such that $C_H(R_0)$ is a Z-group. Let p be a prime. Then $[H, R]$ has p-length one.

Proof. Suppose otherwise and let G be a counterexample of minimal order. Let $r = |R_0|$. Then $[H, R]$ does not have p-length one and clearly $p \neq r$. We claim:

$$(3.6) \qquad\qquad H = [H, R].$$

Suppose that $[H, R] \neq H$. Recall that $[H, R] \lhd \langle H, R \rangle = G$, and therefore $[H, R]$ is a normal Hall subgroup of $[H, R]R$. The group $[H, R]R$ has smaller order than G, and so, by minimality of $|G|$, $[[H, R], R]$ has p-length one. But, by Proposition 1.6(b), $[H, R] = [[H, R], R]$, which, by assumption, does not have p-length one. This proves (3.6).

Suppose $X \lhd H$ is a nontrivial R-invariant subgroup of H. Then R is an operator group on H/X and $|H/X| < |H|$. By Proposition 1.5(d), $C_{H/X}(R_0) = C_H(R_0)X/X$. Therefore, by minimality of G, $[H/X, R]$ has p-length one. But, by (3.6), $[H/X, R] = [H, R]/X = H/X$, so we have:

$$(3.7) \quad H/X \text{ has } p\text{-length one whenever } 1 \neq X \lhd H \text{ and } X^R = X.$$

In particular, if $\mathcal{O}_{p'}(H) \neq 1$, then $H/\mathcal{O}_{p'}(H)$ has p-length one, and hence, by Lemma 1.21(b), H has p-length one, a contradiction. Thus

$$(3.8) \qquad\qquad \mathcal{O}_{p'}(H) = 1.$$

Let $V = F(H)$. Our first main goal is to show that V is elementary abelian.

Since $\mathcal{O}_{p'}(H) = 1$, $F(H)$ is a p-group, so

$$V = F(H) = \mathcal{O}_p(H).$$

Let W be the preimage in H of $\mathcal{O}_{p'}(H/\Phi(V))$. Now $W/\Phi(V)$ and $V/\Phi(V)$ are normal subgroups of $H/\Phi(V)$ of relatively prime orders and hence centralize each other. Thus $[W, V] \subseteq \Phi(V)$. Therefore every p'-element of W centralizes $V/\Phi(V)$ and, by Theorem 1.8, centralizes V. But, by Proposition 1.3, $C_H(V) \subseteq V$ and hence W is a p-group. Thus $\mathcal{O}_{p'}(H/\Phi(V)) = 1$.

Suppose that $\Phi(V) \neq 1$. As $\Phi(V)$ char V char $H \lhd G$, we know that $\Phi(V) \lhd G$ and, by (3.7) (with $X = \Phi(V)$), $H/\Phi(V)$ has p-length one. But, since $\mathcal{O}_{p'}(H/\Phi(V)) = 1$, Lemma 1.21(c) implies that H has p-length one, contrary to our hypothesis. Thus $\Phi(V) = 1$ and, by Lemma 1.7,

$$(3.9) \qquad \begin{array}{c} V = F(H) = \mathcal{O}_p(H) \text{ and} \\ V \text{ is an elementary abelian } p\text{-group.} \end{array}$$

By (3.9) and Proposition 1.3,

$$(3.10) \qquad C_H(V) = V.$$

If H contains two minimal normal subgroups of G, then, by (3.7) and Lemma 1.21(e), H has p-length one, a contradiction. Hence

(3.11) V contains only one minimal normal subgroup of G.

Now let U be the preimage of $F(H/V)$ in H. Since $F(H/V)$ is a p'-group, V is a Sylow p-subgroup of U and so, by Proposition 1.5(a), V has an R-invariant complement K in U. Our next main goal is to show that $N_H(K)$ is a complement to V in H and $C_H(K) \subseteq K$.

Since $N_H(K)$ is R-invariant, there exists an R-invariant Sylow p-subgroup P of $N_H(K)$. By the Schur-Zassenhaus Theorem (**G**, Theorem 6.2.1, p. 221), any two complements of V in U are conjugate in U. Applying the Frattini argument, we have

$$(3.12) \qquad H = U N_H(K) = V N_H(K),$$

and clearly VP is a Sylow p-subgroup of H.

Suppose that $[K, P] = 1$. Then the image of P in H/V centralizes the image of K in H/V. Thus, by Proposition 1.3,

$$PV/V \subseteq C_{H/V}(F(H/V)) \subseteq F(H/V) = KV/V,$$

which implies that $P \subseteq V$, and hence V is a Sylow p-subgroup of H. Consequently H has p-length one, a contradiction. Thus

$$(3.13) \qquad [K, P] \neq 1.$$

By Proposition 1.6(d), $V = C_V(K) \times [V, K]$ and, since K is R-invariant, both $C_V(K) \triangleleft G$ and $[V, K] \triangleleft G$. Therefore (3.11) implies that one of these subgroups is trivial. But, by (3.10), $C_V(K) \neq V$. Thus

$$(3.14) \qquad [V, K] = V, \ C_V(K) = 1, \text{ and } V \cap N_H(K) = 1.$$

Consequently $P \cap V = 1$. Moreover, $N_H(K) \cong H/V$ and $K \cong F(H/V)$, and therefore

$$(3.15) \qquad K - F(N_H(K)).$$

Furthermore, by Proposition 1.3,

$$(3.16) \qquad C_H(K) \subseteq C_{N_H(K)}(K) \subseteq K.$$

Next we study the action of R_0 on H. We show that $[K, R_0] \neq 1$, $|C_V(R_0)| = p$, and $C_P(R_0) = 1$. Assume first that $[K, R_0] = 1$. By (3.15) and Proposition 1.4, $[N_H(K), R_0] = 1$. Thus $N_H(K) \subseteq C_H(R_0)$, and therefore, by hypothesis, K is cyclic. By (3.16), since K is abelian, $C_H(K) = C_{N_H(K)}(K) = K$. Also, since K is cyclic, Aut K is abelian. Since the elements of $N_H(K)$ and R induce automorphisms of K, this implies $[N_H(K), R] \subseteq C_H(K) \subseteq K$. Moreover, since

$$[N_H(K), R] \cong [H/V, R] = [H, R]/V = H/V \cong N_H(K),$$

we have

$$P \subseteq N_H(K) = [N_H(K), R] \subseteq K.$$

But $(|P|, |K|) = 1$, so this implies $P = 1$, contrary to (3.13). Thus we can conclude that

(3.17) $[K, R_0] \neq 1.$

Consider the action of KR_0 on V. Since $C_{KR_0}(V) \lhd KR_0$, we know that $C_K(V) = K \cap C_{KR_0}(V)$ and $C_{R_0}(V) = R_0 \cap C_{KR_0}(V)$ are Hall $\pi(K)$- and Hall $\pi(R_0)$-subgroups of $C_{KR_0}(V)$, respectively. By (3.10),

$$C_K(V) = K \cap C_H(V) \subseteq K \cap V = 1,$$

so $C_{KR_0}(V) = C_{R_0}(V)$. Since R_0 is cyclic of prime order, we know that either $C_{R_0}(V) = 1$ or $C_{KR_0}(V) = R_0$. But in the latter case $R_0 \lhd KR_0$ and hence $[K, R_0] = 1$, contrary to (3.17). Thus

(3.18) $C_{KR_0}(V) = 1.$

Now suppose that $C_V(R_0) = 1$. By (3.18), KR_0 acts faithfully on V. Consequently, by Theorem 3.4, $[K, R_0] = 1$, contrary to (3.17). Therefore $|C_V(R_0)| > 1$.

Since V is elementary abelian and $C_V(R_0) \subseteq C_H(R_0)$, which, by hypothesis, is a Z-group,

(3.19) $|C_V(R_0)| = p.$

Now $C_P(R_0) \subseteq C_H(R_0)$ and $C_V(R_0) \lhd C_H(R_0)$. Since $C_V(R_0)$ is cyclic of order p, $C_V(R_0)$ has no automorphisms of p-power order. But $C_P(R_0)$ is a p-group, and therefore $C_P(R_0)$ centralizes $C_V(R_0)$. Furthermore, $C_P(R_0) \cap C_V(R_0) \subseteq P \cap V = 1$, so $C_P(R_0) \times C_V(R_0) \subseteq C_H(R_0)$. Since $C_H(R_0)$ is a Z-group,

(3.20) $C_P(R_0) = 1.$

Moreover, by Proposition 1.6(a),

(3.21) $P = [P, R_0]C_P(R_0) = [P, R_0].$

Our next main goal is to determine the structure of G more precisely by showing that $H = VKP$ and $R = R_0$ and determining the structure of K. Suppose that $X = X^{PR_0} \subseteq K$ and $VXPR_0 \neq G$. Then $O_{p'}(VXP)$ and V are both normal subgroups of VXP and have relatively prime orders. Thus

$\mathcal{O}_{p'}(VXP)$ centralizes V. By (3.10), $\mathcal{O}_{p'}(VXP) = 1$ and, by minimality of G, we know that $[VXP, R_0]$ has p-length one. Clearly

$$\mathcal{O}_{p'}([VXP, R_0]) \text{ char } [VXP, R_0] \lhd VXP.$$

Therefore $\mathcal{O}_{p'}([VXP, R_0]) \subseteq \mathcal{O}_{p'}(VXP) = 1$, and hence $[VXP, R_0]$ has a normal Sylow p-subgroup, $\mathcal{O}_p([VXP, R_0])$. By (3.21), $P \subseteq \mathcal{O}_p([VXP, R_0])$, and consequently

$$[X, P] \subseteq [X, \mathcal{O}_p([VXP, R_0])] \subseteq \mathcal{O}_p([VXP, R_0]).$$

But $[X, P] \subseteq X$ and p does not divide the order of X, so $[X, P] = 1$.

If $G \neq VKPR_0$ then the case of $X = K$ yields $[K, P] = 1$, contrary to (3.13). Thus $G = VKPR_0$ and hence

(3.22) $$H = VKP, \quad RF = R_0, \text{ and}$$

(3.23) $$[X, P] = 1 \text{ whenever } X = X^{PR} \subset K.$$

Suppose that $K \neq [K, P]$. Since K and P are both PR-invariant, $[K, P]$ is also PR-invariant. Thus, by (3.23), we have $[[K, P], P] = 1$. By Proposition 1.6(b), however, $[K, P] = [[K, P], P] = 1$, contrary to (3.13). Thus

(3.24) $$K = [K, P].$$

Since $K = F(N_H(K))$, we know that K is nilpotent. If $|K|$ is divisible by more than one prime, then, by (3.23), each Sylow subgroup of K is centralized by P. Thus $[K, P] = 1$, which contradicts (3.13). Therefore K is a q-group for some prime q different from both p and r.

Now, by equations (3.16) and (3.23) and **G**, Theorem 5.3.7, p. 181, (taking $A = PR$, $1 \neq \psi \in P$, and letting P be K), we have

(3.25) $$K \text{ is a special } q\text{-group and } C_{K/K'}(P) = 1.$$

(The second statement follows because PR is irreducible on K/K' and $C_{K/K'}(P) \neq K/K'$.) Furthermore, by Theorem 1.13,

(3.26) $$K \text{ has exponent } q.$$

Now consider the action of PR on K. We wish to show that $C_K(R)$ has order q and intersects K' and $[K, R]$ trivially. Since $C_{PR}(K) \lhd PR$, we know that $C_P(K) = P \cap C_{PR}(K)$ and $C_R(K) = R \cap C_{PR}(K)$ are Sylow p- and Sylow r-subgroups of $C_{PR}(K)$, respectively. But, by (3.16),

(3.27) $$C_P(K) = P \cap C_H(K) \subseteq P \cap K = 1.$$

Furthermore, since R is cyclic of prime order and $[K, R] \neq 1$, by (3.17), we have $C_R(K) = 1$. Thus

(3.28) $$C_{PR}(K) = 1.$$

Now K is special, by (3.25), so $K' = \Phi(K)$ and hence, by (3.28) and Theorem 1.8,

$$(3.29) \qquad\qquad C_{PR}(K/K') = 1.$$

Suppose as well that $C_{K/K'}(R) = 1$. Since PR acts faithfully on K/K', Theorem 3.4 and equation (3.29) yield

$$[P, R] = 1,$$

contrary to (3.20). Hence

$$(3.30) \qquad\qquad C_{K/K'}(R) \neq 1$$

and, by Proposition 1.5(d), $C_K(R) \not\subseteq K'$.

Suppose q^2 divides $|C_K(R)|$. Then $C_K(R)$ has an abelian subgroup of order q^2. By (3.26), K has exponent q, so this subgroup must be elementary abelian, contrary to the hypothesis that $C_H(R)$ is a Z-group. Consequently

$$(3.31) \qquad\qquad |C_K(R)| = q \text{ and } C_K(R) \cap K' = 1.$$

By Proposition 1.5(d) and Proposition 1.6(d),

$$K/K' = C_{K/K'}(R) \times [K/K', R] = C_K(R)K'/K' \times [K, R]K'/K'.$$

Therefore

$$(3.32) \qquad\qquad K \neq [K, R].$$

Furthermore, $C_K(R) \cap [K, R] \subseteq K'$, so, by (3.31)

$$(3.33) \qquad\qquad C_{[K,R]}(R) = C_K(R) \cap [K, R] = 1.$$

Our next goal is to show that K is elementary abelian. First note that, by Lemma 3.1, $[K, R]R$ is a Frobenius group with kernel $[K, R]$ and complement R.

Now, by (3.18), $[K, R]R$ acts faithfully on V and, by (3.19), $C_V(R)$ is one-dimensional. Thus we can now apply Theorem 3.5 to the group $[K, R]R$ to conclude that

$$(3.34) \qquad\qquad [K, R] \text{ is abelian.}$$

Suppose that $[K, R]$ is P-invariant. Since, by (3.32), $[K, R]$ is a proper subgroup of K, we know from (3.23) that $[[K, R], P] = 1$ and therefore $[K, R] \subseteq C_K(P)$. But, by (3.25) and Proposition 1.5(d),

$$C_K(P)K'/K' = C_{K/K'}(P) = 1.$$

Thus $[K, R] \subseteq C_K(P) \subseteq K'$. This implies that $[K/K', R] = 1$. By (3.25) and Theorem 1.8, this yields $[K, R] = 1$, contrary to (3.17). Hence $[K, R]$ is not P-invariant. Therefore there exists an $x \in P$ such that

$$(3.35) \qquad\qquad [K, R] \neq [K, R]^x.$$

Now, by Proposition 1.6(a), $K = C_K(R)[K, R]$, so the subgroup $[K, R]$ and its conjugates are abelian subgroups each of prime index q in K. Let x be as in (3.35). Then $K = [K, R][K, R]^x$, $[K, R] \cap [K, R]^x \subseteq Z(K)$, and $|K : [K, R] \cap [K, R]^x| = q^2$. Thus $|K : Z(K)| \leq q^2$. If $|K : Z(K)| = q^2$, then $Z(K) = K' = \Phi(K)$ and PR acts faithfully on the two-dimensional vector space K/K' over $\mathbf{F}_q$. Hence, by Theorem 2.6(a), PR is abelian. Thus $[P, R] = 1$, contrary to (3.21). Therefore $|K : Z(K)| \leq q$ and K is abelian. By (3.26),

(3.36) K is elementary abelian.

Now we will obtain a contradiction by studying the action of K on V. By (3.28), PR acts faithfully on K. Again, applying Theorem 2.6, we can conclude that

(3.37) $|K| > q^2$.

Let $K_1, K_2, \ldots, K_n$ be all of the subgroups of index q in K such that $C_V(K_i) = V_i \neq 1$. Since K is abelian and not cyclic, by Proposition 1.16, the subgroups V_i generate V. We claim:

$$V = V_1 \times V_2 \times \cdots \times V_n.$$

Suppose (renumbering if necessary) that $W = V_1 \times V_2 \times \cdots \times V_m$ is a maximal direct product. Then $V_i \cap W \neq 1$ for any i, $1 \leq i \leq n$. Thus

(3.38) $C_W(K_i) = C_{V_1}(K_i) \times C_{V_2}(K_i) \times \cdots \times C_{V_m}(K_i) \neq 1.$

Take $j \neq i$. Then $K_i \cup K_j$ generates K and hence, by (3.14),

$$C_{V_i}(K_j) = V_i \cap C_V(K_j) = C_V(K_i) \cap C_V(K_j) \subseteq C_V(K) = 1.$$

This, together with (3.38), implies that $1 \leq i \leq m$. Hence $m = n$ and $W = V$, as desired.

By (3.11), V contains only one minimal normal subgroup of G. Clearly RP permutes the subgroups V_i by conjugation. Since distinct orbits of RP on $\{V_1, V_2, \ldots, V_n\}$ would generate normal subgroups of G containing distinct minimal normal subgroups of G, RP must be transitive on $\{V_1, V_2, \ldots, V_n\}$.

Suppose that $V_i^R = V_i$ for all i, $1 \leq i \leq n$. Then $R \subseteq N_{PR}(V_i)$ for every i, and the intersection of all these normalizers is normal in PR. Thus, by (3.21), for every j,

$$P = [P, R] \subseteq \left[P, \bigcap_{i=1}^{n} N_{PR}(V_i)\right] \subseteq \bigcap_{i=1}^{n} N_{PR}(V_i) \subseteq N_{PR}(V_j).$$

Thus $V_i^P = V_i$ for all i. But then V_i is G-invariant and hence $n = 1$. But this contradicts the fact that K acts faithfully on V and $|K| > q^2$ (so $K_1 \neq 1$).

Let $\{V_1, V_2, \ldots, V_r\}$ be a nontrivial R-orbit, and suppose that x is a generator of R. Then for any element $v \in V_1$, writing the operation in V additively, we have

$$v + vx + vx^2 + \cdots + vx^{r-1} \in C_{V_1 \times V_2 \times \cdots \times V_r}(R),$$

and hence the projection map

$$\pi_1 \colon V_1 \times V_2 \times \cdots \times V_r \to V_1$$

takes $C_{V_1 \times V_2 \times \cdots \times V_r}(R)$ onto V_1. But now (3.19) implies that $|V_1| = p$ and

$$C_{V_{r+1} \times V_{r+2} \times \cdots \times V_n}(R) = 1.$$

It follows that the action of R on $\{V_1, V_2, \ldots, V_n\}$ cannot have a second orbit of length r, and hence $V_i{}^R = V_i$ for all i, $r + 1 \le i \le n$.

Suppose that $i > r$. Both K and R induce automorphisms of V_i and, since V_i is cyclic of prime order, its automorphism group is abelian. Thus $[K, R] \subseteq C_K(V_i) = K_i$ for each $i > r$ and, since $[K, R]$ has index q in K, this implies $[K, R] = K_i$ for all $i > r$. Thus either $n = r$ or $n = r + 1$.

Now suppose $n = r$. Since P permutes the subgroups V_i and the order of P is relatively prime to r, some orbit of P has length one. Thus, for some i, P fixes V_i. But then K and P induce automorphisms of V_i and, since $|V_i| = |V_1| = p$, this implies $[K, P]$ centralizes V_i. But, by (3.24), $K = [K, P]$, so K centralizes V_i. This contradicts (3.14).

Finally, suppose that $n = r + 1$. Then n is even. But, since PR acts transitively on the set of subgroups $\{V_i\}$, $n = |PR : N_{PR}(V_{r+1})|$, which is odd. This final contradiction concludes the proof. $\square$

The following theorem (without restriction to groups of odd order) was published by G. Higman in 1957. In his doctoral dissertation, J. G. Thompson generalized the theorem to encompass all finite groups. This more general result is proven in **G**, Theorem 10.2.1, p. 337.

Theorem 3.7. Let $G = KR$ be a solvable group of odd order such that $K \lhd G$ and R is a complement of K of prime order p. Suppose that $C_K(R) = 1$. Then K is nilpotent.

Proof. Let S be a Sylow p-subgroup of G that contains R and set $P = S \cap K$. If p divides $|K|$, then $P \ne 1$, and hence $P \cap Z(S) \ne 1$, which contradicts the hypothesis that $C_K(R) = 1$. Therefore

$$(3.39) \qquad\qquad (|K|, |R|) = 1.$$

We may assume that $K \ne 1$. Take $L \lhd G$ maximal subject to $L \subset K$. Then $C_L(R) \subseteq C_K(R) = 1$. Applying induction to LR in place of G, we see that L is nilpotent. Since $L \lhd G$, it follows that $L \lhd K$ and

$$(3.40) \qquad\qquad L \subseteq F(K).$$

Let $\overline{G} = G/L$, $\overline{K} = K/L$, and $\overline{R} = RL/L$. By Lemmas 3.1 and 3.2,

(3.41)
$$G \text{ and } \overline{G} \text{ are Frobenius groups with kernels } K \text{ and } \overline{K}$$
$$\text{and complements } R \text{ and } \overline{R}, \text{ respectively.}$$

We wish to show that K is nilpotent or, equivalently, that $K \subseteq F(K)$. By Proposition 1.2, it suffices to show that K centralizes every chief factor X/Y of G for which $X \subseteq K$. Take any such chief factor $V = X/Y$ of G. By (3.40) and Proposition 1.2, L centralizes V, whence conjugation by G on X induces irreducible actions of G and of $\overline{G}$ on V. Thus we must show that $\overline{K}$ centralizes V.

Since G is solvable, $\overline{K}$ and V are elementary abelian groups. If $\overline{K}$ and V are q-groups for the same prime q, then $\overline{K}$ centralizes V, as desired, because $\overline{G}$ acts irreducibly on V and $\overline{K} \subseteq \mathcal{O}_q(\overline{G})$ (**G**, Theorem 3.1.3, p. 62). Therefore we may assume that $|\overline{K}|$ and $|V|$ are relatively prime. Then, by (3.39), $|V|$ is relatively prime to $|\overline{G}|$. Since

$$C_V(\overline{R}) = C_V(R) = C_{X/Y}(R) = C_X(R)Y/Y = 1,$$

(3.41) and Lemma 3.3 imply that $\overline{K}$ acts trivially on V, as desired.

This completes the proof of Theorem 3.7.   □

Theorem 3.8. Let $G = KR$ be a solvable group of odd order with $K \lhd G$ and suppose that

(1) $(|R|, |K|) = 1$;
(2) $C_K(x) = C_K(R)$ for all $x \in R^\#$; and
(3) $C_{F(K)}(R) = 1$.

Then $[K, R] \subseteq F(K)$.

Remark. The counterexamples for $\mathbf{F} = \mathbf{F}_p$ mentioned in the remark following Theorem 3.4 yield counterexamples here if one allows KR to have even order. Let KR be the semidirect product of V by G, $K = V\mathcal{O}_2(G)$, and take R to be any group of order three in G. Here, $F(K) = V$.

Proof. Proceed by induction on $|RK|$.

Let $\overline{K} = K/F(K)$. Suppose that R centralizes every Sylow subgroup of $F(\overline{K})$. Then clearly R centralizes $F(\overline{K})$, and hence, by Proposition 1.4, R centralizes $\overline{K}$. But then $[K, R] \subseteq F(K)$ and we are done. Thus we can assume that R does not centralize a Sylow p-subgroup $\overline{P}$ of $F(\overline{K})$ for some prime p.

Let P be the preimage in K of $\overline{P}$. Since $F(K)$ char K and $\overline{P}$ char $F(\overline{K})$, we know that $P \lhd K$. Thus $F(P) \subseteq F(K)$. On the other hand, $F(K) \subseteq P$ and $F(K)$ char K imply that $F(K) \subseteq F(P)$. Thus $F(P) = F(K)$. Then $C_{F(P)}(R) = C_{F(K)}(R) = 1$, and hence (3) holds with K replaced by P. Clearly (1) and (2) also hold with K replaced by P and, consequently, if $P \neq K$, by induction, $[P, R] \subseteq F(P) = F(K)$. But then R centralizes $\overline{P}$, contrary to our choice of P. Thus $K = P$ and $\overline{K}$ is a p-group.

Now suppose that R_0 is a subgroup of R of prime order. Clearly (1) and (2) hold with R replaced by R_0. Furthermore,

$$C_{F(K)}(R_0) = F(K) \cap C_K(R_0) = F(K) \cap C_K(R) = C_{F(K)}(R) = 1,$$

so (3) also holds with R replaced by R_0. If $R_0 \neq R$, induction yields $[K, R_0] \subseteq F(K)$. But then, by Proposition 1.5(d) and (2), we have

$$\overline{K} = C_{\overline{K}}(R_0) = C_K(R_0)F(K)/F(K) = C_K(R)F(K)/F(K) = C_{\overline{K}}(R).$$

In this case $[K, R] \subseteq F(K)$ and we are done. Thus we can assume that $R = R_0$, i.e., R has prime order.

Now let $W = U/V$ be any chief factor of KR with $U \subseteq F(K)$. Since $F(K)$ is nilpotent, W is an elementary abelian q-group for some prime q, and we can consider W as a vector space over $\mathbf{F}_q$. Furthermore, W is irreducible as a module for KR and hence an irreducible and faithful module for $KR/C_{KR}(W)$.

By Proposition 1.5(d),

$$C_W(R) = C_U(R)V/V \subseteq C_{F(K)}(R)V/V = 1.$$

Since $(|K|, |R|) = 1$ and $C_{KR}(W) \lhd KR$, it follows that $C_{KR}(W) \subseteq K$ and therefore $C_{KR}(W) = C_K(W)$. Furthermore, by Proposition 1.2, $F(K) \subseteq C_K(W)$, and hence $K/C_K(W)$ is a p-group.

If $q = p$, then $C_W(K/C_K(W)) \neq 1$ by **G**, Lemma 2.6.3, p. 31. Since $C_W(K/C_K(W))$ is a submodule of W, the irreducibility of W implies that $W = C_W(K/C_K(W))$ or, in other words, $K = C_K(W)$ and, in this case, $[K, R] \subseteq C_K(W)$.

If $q \neq p$, then, by Theorem 3.4, $[RK/C_K(W), K/C_K(W)] = 1$ and hence $[K, R] \subseteq C_K(W)$.

Thus, in either case, we have $[K, R] \subseteq C_K(W)$. Finally, by Proposition 1.2, $[K, R] \subseteq F(K)$, as desired. $\square$

Recall from Section 1 that an operator group A acts *regularly* on a group G if $C_G(\alpha) = 1$ for all $\alpha \in A^{\#}$.

Proposition 3.9. Suppose that p is an odd prime, H is a p'-group, and R is a p-group that acts regularly on H. Then R is cyclic.

Proof. We can assume that $H \neq 1$. Let q be a prime divisor of $|H|$. By Proposition 1.5(a), R fixes, and hence acts regularly on, some Sylow q-subgroup of H. By **G**, Theorem 5.3.14, p. 186, R is cyclic. $\square$

Theorem 3.10. Let $G = KR$ be a solvable Frobenius group with Frobenius kernel K and Frobenius complement R. Suppose that G acts on a nonidentity nilpotent group M such that

(1) $(|G|, |M|) = 1$;
(2) $C_M(K) = 1$; and
(3) $C_M(x) = C_M(R)$ for all $x \in R^{\#}$.

Then:

    (a) R is cyclic of prime order, say, p;

    (b) $|M| = |C_M(R)|^p$; and

    (c) if $C_M(R)$ is cyclic, then $K' \subseteq C_K(M)$.

Proof. We proceed by induction on $|G| + |M|$.

If R_0 is a proper subgroup of R, then KR_0 is still a Frobenius group and (1), (2), and (3) follow with G replaced by KR_0. Hence, by induction, R_0 is cyclic of prime order. Since R is solvable, P. Hall's Theorem (**G**, Theorem 6.4.1, p. 231) guarantees that R has a Hall π-subgroup for every set of primes π. Consequently, either $|R| = p$, $|R| = p^2$, or $|R| = pq$ for distinct primes p and q. Since KR is a Frobenius group, R acts regularly by conjugation on K. Thus, by Proposition 3.9,

$$R \text{ is cyclic.}$$

We finish the proof by dividing it into two cases.

Case 1. M contains a proper G-invariant normal subgroup.

Let M_0 be a normal subgroup of M chosen to be maximal subject to being G-invariant. Clearly $C_{M_0}(K) = M_0 \cap C_M(K) = 1$ and $C_{M_0}(x) = M_0 \cap C_M(x) = M_0 \cap C_M(R) = C_{M_0}(R)$ for all $x \in R^\#$, and hence KR satisfies (1), (2), and (3), with M replaced by M_0. Thus, by induction,

(3.42)
$$|R| = p, \ |M_0| = |C_{M_0}(R)|^p, \text{ and}$$
$$\text{if } C_M(R) \text{ is cyclic, then } K' \subseteq C_K(M_0).$$

This yields (a).

Now KR acts on M/M_0 and $(|G|, |M/M_0|) = 1$. Furthermore, by Proposition 1.5(d),

$$C_{M/M_0}(K) = C_M(K)M_0/M_0 = M_0/M_0 = 1, \text{ and}$$
$$C_{M/M_0}(x) = C_M(x)M_0/M_0 = C_M(R)M_0/M_0 = C_{M/M_0}(R),$$

for all $x \in R^\#$. Thus KR satisfies (1), (2), and (3), with M replaced by M/M_0. By induction, we get

$$|M/M_0| = |C_{M/M_0}(R)|^p \text{ and}$$
$$\text{if } C_M(R) \text{ is cyclic, then } K' \subseteq C_K(M/M_0).$$

Now Proposition 1.5(d) yields

$$|M/M_0| = |C_{M/M_0}(R)|^p = |C_M(R)M_0/M_0|^p = \frac{|C_M(R)|^p}{|C_{M_0}(R)|^p}.$$

Therefore, by (3.42),

$$|M| = |C_M(R)|^p.$$

Thus we have (b).

Finally, combining (3.42) and (3.43) with Lemma 1.9, we see that if $C_M(R)$ is cyclic, then

$$K' \subseteq C_K(M/M_0) \cap C_K(M_0) \subseteq C_K(M).$$

This is (c).

Case 2. M contains no proper G-invariant normal subgroups.

Since M is a minimal normal subgroup of the solvable group MG, M is an elementary abelian r-group for some prime r. Hence we can regard M as a vector space over $\mathbf{F}_r$. Thus (c) follows from (a) and Theorem 3.5. Therefore it suffices to prove (a) and (b).

By hypothesis, M is irreducible as a $\mathbf{F}_r G$-module. By Proposition 2.1, $V = M \otimes_\mathbf{K} \overline{\mathbf{K}}$ is an irreducible $\overline{\mathbf{K}}G$-module for the algebraic closure $\overline{\mathbf{K}}$ of $\mathbf{K} = \mathrm{Hom}_{\mathbf{F}_r G}(M, M)$.

Suppose that $K_0 \lhd KR$ is chosen to be minimal subject to $1 \subset K_0 \subseteq K$. Since $C_M(K_0)$ is a G-invariant subgroup of M, either $C_M(K_0) = 1$ or $C_M(K_0) = M$. In the first case, (1), (2), and (3) are satisfied with G replaced by $K_0 R$. In the second case, (1), (2), and (3) are satisfied with G replaced by KR/K_0. Thus, if $K_0 \neq K$, induction yields (a) and (b).

Therefore we can assume that K is a minimal normal subgroup of KR. Since KR is solvable, this implies that K is an elementary abelian q-group for some prime q.

Since KR is a Frobenius group and $C_{KR}(V) \lhd KR$, we have, by (2) and Lemma 3.2, $C_{KR}(V) = C_K(V) = 1$. Thus KR acts faithfully on V.

We are now in a position to apply **G**, Theorem 3.4.3, p. 73 and Clifford's Theorem (**G**, Theorem 3.4.1, p. 70) to conclude that

$$V = \bigoplus_{x \in R}(Wx),$$

where W is a Wedderburn component of V with respect to K. Consequently

(3.43)
$$C_V(R) = \left\{ \sum_{x \in R} wx \mid w \in W \right\} \text{ and}$$
$$\dim(C_V(R)) = \dim(W) = \frac{1}{|R|}\dim(V).$$

Thus

(3.44)
$$|C_M(R)|^{|R|} = |W|^{|R|} = |M|.$$

Now let P be a subgroup of R of order p and let $x_1, x_2, \ldots, x_s$ be a set of left coset representatives for P in R. Then

$$V = \bigoplus_{\substack{g \in P \\ 1 \le i \le s}}(Wx_i g) = \bigoplus_{g \in P}\left(\bigoplus_{1 \le i \le s} Wx_i\right)g,$$

and hence
$$C_V(P) = \left\{ \sum_{g \in P} wg \,\middle|\, w \in \bigoplus_{1 \leq i \leq s} W x_i \right\},$$
which implies that
$$\dim C_V(P) = \dim \bigoplus_{1 \leq i \leq s} W x_i = |R : P| \dim W.$$

But, by (3), $\dim C_V(P) = \dim C_V(R)$. Thus (3.43) implies $|R : P| = 1$. This gives us (a). Finally, (3.44) now yields
$$|C_M(R)|^p = |M|,$$
and we have (b). $\square$

4. p-Groups of Small Rank

The main purpose of this section is to investigate solvable groups G of odd order that contain no elementary abelian p-subgroups of rank three, either for a single prime p or for all primes p. The general structure of these groups is described in Theorem 4.18 and Theorem 4.20, respectively. The p-groups in this family that are most important for our purposes are those that are determined in a result of Blackburn, Theorem 4.16. We shall see later that every Sylow p-subgroup of rank two in a minimal counterexample to the Odd Order Theorem is a p-group of this type.

If A is an abelian p-group, let $\mathrm{m}(A)$ be the minimal number of generators of A (as in **G**, p. 8). Thus, if A is a p-group for some prime p, then $|\Omega_1(A)| = p^{\mathrm{m}(A)}$. For each prime q, define the q-rank of G by
$$\mathrm{r}_q(G) = \max \{ \mathrm{m}(A) \mid A \text{ is an abelian } q\text{-subgroup of } G \}.$$
Define the *rank* of G by
$$\mathrm{r}(G) = \max \{ \mathrm{r}_q(G) \mid q \text{ is a prime} \}.$$
These are called the *q-depth* and *depth* of G in **G**, pp. 188-189.

For any prime p, natural number n, and p-subgroup R of G, define (as in **G**, p. 17 and pp. 288-289)
$$\mho^n(R) = \left\langle x^{p^n} \mid x \in R \right\rangle,$$
$$\mathrm{SCN}(R) = \{ A \mid A \triangleleft R \text{ and } C_R(A) = A \}, \text{ and}$$
$$\mathrm{SCN}_n(R) = \{ A \mid A \in \mathrm{SCN}(R) \text{ and } \mathrm{m}(A) \geq n \}.$$

We say that a group G is *metacyclic* if G possesses a cyclic normal subgroup N with G/N cyclic. We say that G is a *central product* of subgroups $G_1, \ldots, G_n$ and write
$$G = G_1 \circ \cdots \circ G_n,$$
if $G_i \triangleleft G$ for each i, G_i centralizes G_j for each $i \neq j$, and $G = G_1 \cdots G_n$.

We will use the symbol $\mathrm{cl}(G)$ to denote the nilpotence class of G.

Lemma 4.1. If G is a group and $G/Z(G)$ is cyclic, then G is abelian.

Proof. G, Theorem 1.3.4, p. 11. □

Lemma 4.2. Suppose that G is a group, x, $y \in G$, and $[x, y] \in Z(G)$. Then for all $n \geq 1$,

(a) $[x^n, y] = [x, y]^n = [x, y^n]$, and
(b) $(xy)^n = x^n y^n [y, x]^{\binom{n}{2}}$,

where $\binom{n}{2}$ denotes the usual binomial coefficient.

Proof. G, Lemma 2.2.2, p. 19. □

Proposition 4.3. Suppose that p is an odd prime and R is a p-group. Assume

(1) $\mathrm{cl}(R) \leq 2$, or
(2) $p > 3$ and $\mathrm{cl}(R) \leq 3$.

Define a mapping ϕ of R into R by $\phi(x) = x^p$. Then

(a) $\Omega_1(R)$ has exponent 1 or p, and
(b) if $R' \subseteq \Omega_1(R)$, then ϕ is a homomorphism.

Proof. This result follows immediately from Philip Hall's theory of regular p-groups, since the class of R is less than p. (See [19, pp. 183–187] or [17, pp. 321–326].) However, since this topic is not treated in **G**, we provide a proof.

For any elements u, v, and w of a group,

(4.1) $\quad [uv, w] = [u, w][u, w, v][v, w]$ and $[u, vw] = [u, w][u, v][u, v, w]$,

as can be verified by direct calculation. In particular, for any natural number n,

$$[u^{n+1}, w] = [u, w][u, w, u^n][u^n, w].$$

Let u, $w \in R$. Since $\mathrm{cl}(R) \leq 3$, we know $[R, R, R] \subseteq Z(R)$, and therefore, by Lemma 4.2, $[u, w, u^n] = [u, w, u]^n$ for all n, and, by induction on n,

(4.2) $\quad\quad\quad\quad [u^n, w] = [u, w]^n [u, w, u]^{\binom{n}{2}}.$

For each natural number n, define

$$f(n) = \binom{n}{3} \text{ and } g(n) = 2\binom{n}{3} + \binom{n}{2}.$$

Note that for each n,

(4.3) $\quad f(n + 1) = \binom{n}{2} + f(n)$ and $g(n + 1) = 2\binom{n}{2} + \binom{n}{1} + g(n)$.

This follows easily from the Pascal triangle identity

$$\binom{n}{i} + \binom{n}{i+1} = \binom{n+1}{i+1}.$$

Let u, $v \in R$. We claim that for each natural number n

(4.4) $$(uv)^n = u^n v^n [v, u]^{\binom{n}{2}} [v, u, u]^{f(n)} [v, u, v]^{g(n)}.$$

We use induction. The result is obvious for $n = 1$. Assume it is true for $n = k$. Then

$$(uv)^{k+1} = (uv)^k (uv) = u^k v^k [v, u]^{\binom{k}{2}} [v, u, u]^{f(k)} [v, u, v]^{g(k)}(uv)$$

$$= u^k v^k (uv)[v, u]^{\binom{k}{2}} [[v, u]^{\binom{k}{2}}, uv][v, u, u]^{f(k)} [v, u, v]^{g(k)}.$$

By (4.1) and the previous lemma,

$$[[v, u]^{\binom{k}{2}}, uv] = [v, u, uv]^{\binom{k}{2}} = [v, u, u]^{\binom{k}{2}} [v, u, v]^{\binom{k}{2}}.$$

Let

$$w = [v, u]^{\binom{k}{2}} [v, u, u]^{\binom{k}{2}+f(k)} [v, u, v]^{\binom{k}{2}+g(k)}.$$

Then

$$(uv)^{k+1} = u^k v^k (uv)w.$$

Now, by (4.2),

$$v^k uv = uv^k [v^k, u]v = uv^k [v, u]^k [v, u, v]^{\binom{k}{2}} v$$

$$= uv^{k+1} [v, u]^k [[v, u]^k, v][v, u, v]^{\binom{k}{2}}$$

$$= uv^{k+1} [v, u]^k [v, u, v]^{k+\binom{k}{2}}.$$

Hence

$$(uv)^{k+1} = u^k(v^k uv)w = u^{k+1} v^{k+1} [v, u]^k [v, u, v]^{k+\binom{k}{2}} w.$$

By (4.3),

$$(uv)^{k+1} = u^{k+1} v^{k+1} [v, u]^{\binom{k+1}{2}} [v, u, u]^{f(k+1)} [v, u, v]^{g(k+1)}.$$

This completes the proof of (4.4).

We prove (a) by induction on $|R|$. Take elements x, $y \in R$ such that $x^p = y^p = 1$. It is sufficient to show that for any such x and y, we have $(xy)^p = 1$. This is obvious if $\langle x \rangle = R$, so assume that $\langle x \rangle \neq R$.

Let S be a maximal subgroup of R that contains $\langle x \rangle$. Then $S \triangleleft R$ and $\Omega_1(S) \triangleleft R$. By induction, $\Omega_1(S)$ has exponent 1 or p. Since $x \in \Omega_1(S)$,

(4.5) $$[y, x]^p = [y, x, x]^p = [y, x, y]^p = 1.$$

Since p is odd, p divides $\binom{p}{2}$. If $p > 3$, then p divides $f(p)$ and $g(p)$ and, if $p = 3$, then $[R, R, R] = 1$. Hence, by (4.4) and (4.5), $(xy)^p = 1$. This proves (a), and (b) follows similarly from (4.4). $\square$

Proposition 4.4. Suppose p is a prime and R is a p-group. Then

    (a) $\mathrm{SCN}(R)$ is the set of all normal subgroups of R that are maximal with respect to the property of being abelian, and

    (b) if R is a Sylow p-subgroup of a group G and $A \in \mathrm{SCN}(R)$, then $C_G(A) = A \times H$ for some p'-subgroup H of G.

Proof. (a) **G**, Theorem 5.3.12, p. 185. (b) **G**, Theorem 7.6.5, p. 259. $\square$

Lemma 4.5. Suppose p is an odd prime and R is a noncyclic p-group. Then

    (a) R possesses a normal elementary abelian subgroup of order p^2,

    (b) if R possesses a cyclic subgroup of index p, then $\Omega_1(R)$ is elementary abelian of order p^2, and

    (c) $\Omega_1(Z_2(R))$ is a noncyclic group of exponent p.

Proof. (a) **G**, Theorem 5.4.10, p. 199. (b) **G**, Theorem 5.4.4, p. 193 and **G**, Theorem 5.4.3, p. 191. (c) Let $Z = \Omega_1(Z_2(R))$. By (a), R possesses a normal elementary abelian subgroup S of order p^2. Since R is nilpotent

$$[S, R] \subset S.$$

Similarly

$$[S, R, R] \subset [S, R] \text{ if } [S, R] \neq 1.$$

Consequently

$$[S, R, R] = 1, \ [S, R] \subseteq Z(R), \text{ and } S \subseteq Z_2(R).$$

So $S \subseteq Z$. Hence Z is not cyclic. Since $\mathrm{cl}(Z_2(R)) \leq 2$, Z has exponent p by Proposition 4.3(a). $\square$

Proposition 4.6. Suppose p is an odd prime, R is a p-group, and S is a noncyclic normal subgroup of R. Then S contains a normal elementary abelian subgroup of R that has order p^2.

Proof. Let $Z = \Omega_1(Z_2(S))$. By Lemma 4.5, Z has exponent p and order at least p^2. Since $Z \lhd R$ and R is a p-group, Z contains a normal subgroup of R of order p^2 by Lemma 1.22. This subgroup satisfies the conclusion. $\square$

Lemma 4.7. Suppose p is an odd prime and R is a p-group. Then $\mathrm{SCN}_3(R)$ is empty if and only if $\mathrm{r}(R) \leq 2$.

Proof. Obviously, $\mathrm{SCN}_3(R)$ is empty if $\mathrm{r}(R) \leq 2$. The converse is proved in **G**, Theorem 5.4.15, p. 202. $\square$

Remark. Proposition 4.6 and Lemma 4.7 prove the following about a p-group R for an odd prime p: R has an elementary abelian subgroup of order p^2 if and only if R has a normal elementary abelian subgroup of order p^2, and likewise for p^3. The dihedral group D_{16} and the group $D_{16} \times Z_2$ are counterexamples to the corresponding statements for $p = 2$.

Proposition 4.8. Suppose p is a prime, R is a p-group, and $\mathrm{r}(R) \leq 2$.

    (a) If R has exponent p, then $|R| \leq p^3$.

    (b) If $p > 3$, then $\Omega_1(R)$ has exponent one or p.

Proof. (a) Assume that R has exponent p. Take $A \in \mathrm{SCN}(R)$. Since $\mathrm{r}(R) \leq 2$, we have $|A| \leq p^2$. Hence

$$|R/A| = |R/C_R(A)| \leq |\operatorname{Aut} A|_p \leq |\mathbf{GL}(2,p)|_p = p.$$

    (b) Assume this is false and let R be a minimal counterexample. By Proposition 4.3 it suffices to show that $\mathrm{cl}(R) \leq 3$ to obtain a contradiction. By the minimality of R, there exist elements $x, y \in R$ such that $x^p = y^p = 1$, $\langle x \rangle \neq R$, and $\langle x, y \rangle = R$. Let S be a maximal subgroup of R that contains $\langle x \rangle$. Then $S \lhd R$, $x \in \Omega_1(S) \lhd R$, and $R = \langle \Omega_1(S), y \rangle$. By minimality of R, $\Omega_1(S)$ has exponent 1 or p. By (a), $|\Omega_1(S)| \leq p^3$. Thus $|R| \leq p^4$. This implies that $\mathrm{cl}(R) \leq 3$, as desired. $\square$

Lemma 4.9. Suppose p is a prime, $p > 3$, and R is a p-group. Assume that $|\Omega_1(R)| \leq p^2$. Then $|\Omega_1(R/T)| \leq p^2$ for every subgroup $T \lhd R$.

Proof. Suppose R is a minimal counterexample to the lemma and T is a normal subgroup of R chosen to be minimal subject to $|\Omega_1(R/T)| > p^2$.

    Suppose $|T| > p$. Let Z be a subgroup of $T \cap Z(R)$ of order p. By the minimality of T, we then have $|\Omega_1(R/Z)| \leq p^2$. But $|R/Z| < |R|$ and

$$R/T \cong (R/Z)/(T/Z),$$

so, by the minimality of R,

$$|\Omega_1(R/T)| = |\Omega_1((R/Z)/(T/Z))| \leq p^2,$$

which contradicts the choice of T. Thus $|T| = p$.

    If $\mathrm{r}(R/T) > 2$, then, by the minimality of R, we know that R/T is elementary abelian of order p^3. If $\mathrm{r}(R/T) \leq 2$, then, by Proposition 4.8 and the minimality of R, we know R/T has exponent p and order p^3. Consequently, in both cases,

$$(4.6) \qquad\qquad |R| = p^4 \text{ and } R/T \text{ has exponent } p.$$

    By Lemma 4.5 and the hypotheses, $|\Omega_1(R)| = p^2$. Hence $R/\Omega_1(R)$ has order p^2 and is therefore abelian. Define a mapping $\phi\colon R \to R$ by $\phi(x) = x^p$. By (4.6), $\mathrm{cl}(R) \leq 3$ and ϕ maps R into T. By Proposition 4.3, ϕ is a homomorphism. Hence

$$p = |T| \geq |R/\operatorname{Ker}(\phi)| = |R/\Omega_1(R)| = p^4/p^2 = p^2,$$

a contradiction. $\square$

Lemma 4.10. Suppose that p is an odd prime and R is a metacyclic p-group that is not cyclic. Then $\Omega_1(R)$ is elementary abelian of order p^2.

Proof. Take $S \lhd R$ such that S and R/S are cyclic. Let T be the preimage in R of $\Omega_1(R/S)$, so that

$$T/S = \Omega_1(R/S).$$

Then $\Omega_1(R) = \Omega_1(T)$. Now apply Lemma 4.5(b) to the group T. □

Proposition 4.11 (Huppert). Suppose that p is a prime and R is a p-group. Assume $p > 3$ and $|\Omega_1(R)| \le p^2$. Then R is metacyclic.

Proof. This is part of a result of Huppert [17, Satz III.11.6, p. 338]. We prove it by using induction on $|R|$ and applying the previous lemmas.

The result is obvious if R is abelian, so we assume that R is not abelian. Then $1 \subset R' \lhd R$ and hence $R' \cap Z(R) \ne 1$. Let T be a subgroup of order p in $R' \cap Z(R)$, chosen so that $T \subseteq \mho^1(R') \cap Z(R)$ if $\mho^1(R') \ne 1$.

Now $T = \langle z \rangle$ for some element z of order p. Also, by Lemma 4.9, $|\Omega_1(R/T)| \le p^2$. Therefore, by induction, R/T is metacyclic. Hence there exist $a, b \in R$ such that

(4.7) $$\langle a, z \rangle \lhd R \text{ and } R = \langle a, b, z \rangle.$$

Then $R' \subseteq \langle a, z \rangle$. If R/T is cyclic, then $R/Z(R)$ is cyclic and R is abelian by Lemma 4.1. Thus R/T is not cyclic. Therefore $\langle a, z \rangle / \langle z \rangle \ne 1$. Since $\langle a, z \rangle / \langle a^p, z \rangle$ is a normal subgroup of order p in $R/\langle a^p, z \rangle$, it is contained in the center of $R/\langle a^p, z \rangle$. Thus $[a, b] = a^{ip} z^j$ for some integers i, j.

If $1 \subset \mho^1(R')$, then

$$\langle z \rangle = T \subseteq \mho^1(R') \subseteq \mho^1(\langle a, z \rangle) = \mho^1(\langle a \rangle) \subseteq \langle a \rangle.$$

In this case, R is metacyclic by (4.7). Therefore we will assume that $\mho^1(R') = 1$. Since $|\Omega_1(R)| \le p^2$, R' is elementary abelian.

Now

$$a^b = b^{-1}ab = aa^{-1}b^{-1}ab = a[a, b] = a^{1+ip} z^j.$$

Hence a and z centralize a^b and $[a, b]$, and

$$[a, b]^b = (a^{ip} z^j)^b = (a^b)^{ip} z^j = a^{ip}[a, b]^{ip} z^j = a^{ip} z^j = [a, b].$$

Thus $[a, b] \in Z(R)$. Since a, b, and z centralize each other modulo $\langle [a, b] \rangle$,

$$R' = \langle [a, b] \rangle.$$

Let S be a cyclic subgroup of R that is maximal subject to $R' \subseteq S$. Since $R' \subseteq S$, we know $S \lhd R$ and, since R is not cyclic, $S \ne R$. Let S_1 be any subgroup of R that contains S and such that $|S_1/S| = p$. By the maximality of S, S_1 is not cyclic and hence, by Lemma 4.5(b), we know $|\Omega_1(S_1)| = p^2$. Thus $\Omega_1(S_1) = \Omega_1(R)$ and

$$S_1 = \Omega_1(S_1)S = \Omega_1(R)S.$$

Therefore the subgroup S_1 is uniquely determined and R/S has a unique subgroup of order p. Again, by Lemma 4.5, we can conclude that R/S is cyclic, and hence R is metacyclic, as desired. $\square$

Theorem 4.12 (Huppert). Suppose p is an odd prime, R is a metacyclic p-group, and A is a p'-group of operators on R. Then

(a) $[R, A]$ is abelian,

(b) $R = [R, A]C_R(A)$ and $[R, A] \cap C_R(A) = 1$, and

(c) if R is not abelian and A does not act trivially on R, then $[R, A]$ and $C_R(A)$ are nonidentity cyclic groups and $R' \subseteq [R, A]$.

Proof. (a) We use induction on $|R|$. By Proposition 1.6(b), $[R, A] = [R, A, A]$. Therefore we can assume that

$$(4.8) \qquad\qquad R = [R, A].$$

Since R is metacyclic, R' is cyclic. Take a cyclic A-invariant subgroup S of R that is maximal subject to containing R'. Then S is a normal subgroup of R and of the semidirect product RA. Since S is cyclic, $\operatorname{Aut} S$ is abelian. Therefore, by (4.8),

$$R = [R, A] \subseteq (RA)' \subseteq C_{RA}(S).$$

Thus

$$(4.9) \qquad\qquad S \subseteq Z(R).$$

Since $R' \subseteq S$, the group R/S is abelian. Therefore we can regard $\Omega_1(R/S)$ as a vector space over $\mathbf{F}_p$ and as an A-module. By Maschke's Theorem (Theorem 1.20), there exists an A-invariant complement X/S to the A-submodule $\Omega_1(R)S/S$ of $\Omega_1(R/S)$. By Lemma 4.10, $\Omega_1(R)$ is elementary abelian. Thus

$$X \cap \Omega_1(R) = X \cap \Omega_1(R) \cap S = \Omega_1(S).$$

Hence $|\Omega_1(X)| \le |\Omega_1(S)| = p$. Consequently, by Lemma 4.5, X is cyclic. By our maximal choice of S, we have $X = S$. Thus

$$(4.10) \qquad\qquad \Omega_1(R/S) = \Omega_1(R)S/S.$$

By Lemma 4.10, $|\Omega_1(R)| \le p^2$, so, by (4.10),

$$|\Omega_1(R/S)| = |\Omega_1(R)S/S| = |\Omega_1(R)/\Omega_1(R) \cap S| \le |\Omega_1(R)/\Omega_1(S)| \le p.$$

Again, by Lemma 4.5, R/S is cyclic. Thus, by (4.9), we know $R/Z(R)$ is cyclic, and hence R is abelian, as desired.

(b) Let $T = [R, A]$. By (a) and Proposition 1.6, $[T, A] = T$, $R = TC_R(A)$, and $T = [T, A] \times C_T(A) = T \times (T \cap C_R(A))$. Hence $T \cap C_R(A) = 1$.

(c) Let $T = [R, A]$. By (a), $1 \subset T \subset R$, so $C_R(A) \neq 1$. By (b), $T \cap C_R(A) = 1$, and therefore

$$|\Omega_1(R)| \geq |\Omega_1(T)\Omega_1(C_R(A))| = |\Omega_1(T)||\Omega_1(C_R(A))| \geq p^2.$$

By Lemma 4.10, $|\Omega_1(R)| \leq p^2$. Thus $|\Omega_1(T)| = |\Omega_1(C_R(A))| = p$.

By Lemma 4.5, T and $C_R(A)$ are cyclic. Since $T \lhd R$ and $R = TC_R(A)$, we obtain $R' \subseteq T$. This completes the proof of (c) and of the theorem. $\square$

Lemma 4.13. Suppose p is an odd prime, R is a p-group, and q is a prime divisor of $|\operatorname{Aut} R|$. Assume that $\operatorname{SCN}_3(R)$ is empty and $q \neq p$. Then q divides $(p^2 - 1)$ and $q < p$.

Proof. Lemma 4.7 and **G**, Theorem 5.4.15, p. 202. $\square$

Lemma 4.14. In Lemma 4.13, q divides $\frac{1}{2}(p + 1)$ or $\frac{1}{2}(p - 1)$.

Proof. If q is odd use the fact that $p^2 - 1 = 4(\frac{1}{2}(p - 1))(\frac{1}{2}(p + 1))$. $\square$

Lemma 4.15. Suppose S is an extraspecial subgroup of a p-group R and $[S, R] \subseteq S'$. Then $R = SC_R(S)$.

Proof. **G**, Lemma 5.4.6, p. 195. $\square$

Theorem 4.16 (Blackburn). Suppose p is an odd prime, R is a non-identity p-group, and A is a p'-group of automorphisms of R. Assume that $r(R) \leq 2$, $[R, A] = R$, and $|A|$ is odd. Then $p > 3$ and R satisfies one of the following two conditions:

(1) R is abelian, or
(2) $R = R_1 \circ R_2$ for some nonabelian group R_1 of order p^3 and exponent p and some cyclic group R_2 such that $\Omega_1(R_2) = R_1'$.

Proof. Proceed by induction on $|R|$. Clearly $\operatorname{SCN}_3(R)$ is empty and each prime divisor of $|A|$ is odd. By Lemma 4.13, $p > 3$. If $|\Omega_1(R)| \leq p^2$, then Proposition 4.11, Theorem 4.12 and the assumption that $[R, A] = R$, show that R is metacyclic and abelian. In this case R satisfies (1).

For the remainder of the proof, assume that $|\Omega_1(R)| > p^2$. By Proposition 4.8, $\Omega_1(R)$ has exponent p and order p^3. Let $S = \Omega_1(R)$ and $C = C_R(S)$. Since $r(R) = 2$, S is not abelian. Thus S is extraspecial. Furthermore,

$$S' \subseteq \Omega_1(C) \subseteq \Omega_1(R) \cap C = S \cap C = Z(S) = S'.$$

Thus $|\Omega_1(C)| = p$ and, by Lemma 4.5, we have

(4.11) C is cyclic.

Suppose that R centralizes S/S'. Then, by Lemma 4.15,

$$R = CS = SC.$$

By (4.11), C is cyclic and $\Omega_1(C) = S'$. Thus, in this case, (2) holds.

We are left with the case in which R does not centralize S/S'. In this case, let $T = [R, S]$. Then $S' \subset T$ and, since R is nilpotent,

$$T = [R, S] \subset S.$$

Hence $|T| = p^2$. Consequently $|\operatorname{Aut} T| = |\mathbf{GL}(2, p)| = p(p^2 - 1)(p - 1)$. Since $T \lhd R$ and $T \not\subseteq Z(S)$,

$$(4.12) \qquad |R/C_R(T)| = p \text{ and } R = SC_R(T).$$

Take $y \in S - T$. Let B be the group of all automorphisms of S that centralize T and S/T. Suppose $\beta, \gamma \in B$. Let $y^\beta = yt$ and $y^\gamma = yu$. Then $t, u \in T$, and

$$(4.13) \qquad y^{\beta\gamma} = (yt)^\gamma = yut = ytu = y^{\gamma\beta} \text{ and}$$

$$(4.14) \qquad y^{\beta^i} = yt^i \text{ for each } i = 1, 2, 3, \ldots.$$

Since $S = \langle y, T \rangle$, we have $\beta\gamma = \gamma\beta$ and $\beta^p = 1$. Thus B is an elementary abelian p-group. Since $C_R(T)/C$ is isomorphic to a subgroup of B,

$$C_R(T)/C \text{ is an elementary abelian } p\text{-group.}$$

By Maschke's Theorem (Theorem 1.20), TC/C has an A-invariant complement X/C in $C_R(T)/C$. Since $\mathrm{r}(R) \leq 2$, we have $\Omega_1(C_R(T)) = T$, and thus

$$\Omega_1(X) \subseteq \Omega_1(C_R(T)) \cap X \subseteq T \cap X \subseteq C.$$

Since C is cyclic by (4.11), $|\Omega_1(X)| = p$ and, by Lemma 4.5, X is cyclic. Let $X = \langle x \rangle$.

By (4.12), $R = SC_R(T) = SX$. Since both S and T centralize S/S', we know that X does not centralize S/S'. That is,

$$(4.15) \qquad [X, S] \not\subseteq S'.$$

Now $|S/T| = |T/S'| = p$. Therefore, by taking any elements $y \in S - T$ and $z \in T - S'$, we get

$$S/T = \langle yT \rangle \text{ and } T/S' = \langle zS' \rangle.$$

By (4.12) and the hypothesis that $[R, A] = R$, we know $[S, A] \not\subseteq T$. Choose $\alpha \in A$ such that $[S, \alpha] \not\subseteq T$. Then there exist integers i, j, and k such that

$$x^\alpha = x^i,$$
$$y^\alpha \equiv y^j \pmod{T}, \text{ and}$$
$$z^\alpha \equiv z^k \pmod{S'}$$

and

$$y^{\alpha^2} \equiv y^{j^2} \pmod{T}.$$

Since α has odd order, α^2 does not act trivially on S/T, and hence

$$j^2 \not\equiv 1 \pmod{p}.$$

Note that $i \not\equiv 0 \pmod{p}$, because $\langle x^i \rangle = \langle x^\alpha \rangle = \langle x \rangle$.

Now clearly $1 \neq [y, z] \in S' \subseteq \langle x \rangle$ and $[x, y] \in [R, S] = T$. Furthermore, $[x, y] \notin S'$, for otherwise $[X, S] \subseteq S'$, contrary to (4.15). Consequently $[x, y] \in T - S'$. Two applications of Lemma 4.2 now yield

$$[y, z]^i = [y, z]^\alpha = [y^\alpha, z^\alpha] = [y^j, z^k] = [y, z]^{jk}, \text{ and}$$
$$[x, y]^k \equiv [x, y]^\alpha \equiv [x^\alpha, y^\alpha] \equiv [x^i, y^j] \equiv [x, y]^{ij} \pmod{S'}.$$

(In the second equation we have used the fact that $T \subseteq Z(R/S')$, i.e., $[R, T] \subseteq S'$.) Thus

$$jk \equiv i \pmod{p},$$
$$ij \equiv k \pmod{p},$$
$$ij^2 \equiv i \pmod{p}, \text{ and}$$
$$j^2 \equiv 1 \pmod{p},$$

a contradiction. $\square$

Lemma 4.17. Suppose p is an odd prime, R is a p-group, and A is a solvable group of automorphisms of R. Assume that

$$\mathrm{r}(R) \leq 2 \text{ and } |A| \text{ is odd}.$$

Then A' is a p-group.

Proof. We may assume that $R \neq 1$. Take a characteristic subgroup H of R as in Theorem 1.13. Then

(4.16) $C_A(H)$ is a p-group.

Furthermore, H has exponent p and $\mathrm{r}(H) \leq \mathrm{r}(R) \leq 2$. Consequently, by Proposition 4.8,

(4.17) $|H| \leq p^3$.

Let $V = H/\Phi(H)$ and $C = C_A(V)$. Then $C_A(H) \subseteq C$, and $C/C_A(H)$ is isomorphic to a group of automorphisms of H that acts trivially on V. Hence, by Theorem 1.8, $C/C_A(H)$ is a p-group. By (4.16),

(4.18) C is a p-group.

Now

(4.19) A/C is isomorphic to a group of automorphisms of V.

By (4.17), $m(V) \leq 3$. If $m(V) = 3$, then $\Phi(H) = 1$, H is elementary abelian, and

$$3 = m(H) \leq r(R) = 2,$$

a contradiction. If $m(V) = 1$, then, by (4.19), A/C is abelian, and therefore, by (4.18), A' is a p-group.

We are left with the case in which $m(V) = 2$. Here $\operatorname{Aut} V \cong \mathbf{GL}(2,p)$. By (4.19) and Theorem 2.6, $(A/C)'$ is a p-group. Since $(A/C)' = A'C/C$, it follows from (4.18) that $A'C$ is a p-group. Hence A' is a p-group, as desired. $\square$

Theorem 4.18. Suppose that G is a solvable group of odd order, $p \in \pi(G)$, and $r_p(G) \leq 2$. Then:

(a) p is the largest prime divisor of $|G/\mathcal{O}_{p'}(G)|$;

(b) if $p = 3$ or p is the smallest prime divisor of $|G|$, then G has a normal p-complement;

(c) G' has a normal p-complement;

(d) every p'-subgroup of G' is contained in $\mathcal{O}_{p'}(G')$; and

(e) $G/\mathcal{O}_{p',p}(G)$ is an abelian p'-group.

Proof. Note that each desired conclusion is valid for G if it is valid for $G/\mathcal{O}_{p'}(G)$. Therefore we will assume that $\mathcal{O}_{p'}(G) = 1$.

Let $R = \mathcal{O}_p(G)$ and $C = C_G(R)$. Then, by **G**, Theorem 6.3.2, p. 228,

$$C \subseteq R \text{ and } r(R) \leq r_p(G) \leq 2.$$

By Lemma 4.17, $(G/C)'$ is a p-group. Since $(G/C)' = G'C/C$ and C is itself a p-group, $G'C$ is also a p-group. Hence $G'C \subseteq \mathcal{O}_p(G) = R$. Thus G/R is abelian. As $\mathcal{O}_p(G/R) = 1$, we know G/R is a p'-group. By Lemma 4.7, $\mathrm{SCN}_3(R)$ is empty and therefore, by Lemma 4.13, p is the largest prime divisor of $|G|$. Now all parts of the theorem follow easily. $\square$

Corollary 4.19. Suppose that p is a prime, G is a solvable group, and G^* is a normal subgroup of G. Assume that

$$r_p(G^*) \leq 2 \text{ and } |G| \text{ is odd.}$$

Then G' centralizes every chief factor U/V of G such that U/V is a p-group and $U \subseteq G^*$.

Proof. Clearly, as in the proof of Theorem 4.18, we can assume that $\mathcal{O}_{p'}(G^*) = 1$. Let $R = \mathcal{O}_p(G^*)$. By Theorem 4.18, R is a Sylow p-subgroup of G^*. Hence $U \subseteq RV$.

Let $C = C_G(U/V)$. By Lemma 4.17, $(G/C_G(R))'$ is a p-group. Thus $(G/C)'$ is a p-group. As G/C acts faithfully and irreducibly on U/V,

$$1 = \mathcal{O}_p(G/C) \supseteq (G/C)' = G'C/C.$$

Therefore $G' \subseteq C$, as desired. $\square$

Theorem 4.20. Suppose that G is a solvable group of odd order and $r(G) \leq 2$ or $r(F(G)) \leq 2$. Then:

   (a) G' is nilpotent; and
   (b) if S is a Sylow subgroup of G, T is a characteristic subgroup of S, and $T \subseteq S'$, then $T \lhd G$.

Assume $G \neq 1$. Let $\pi(G) = \{p_1, \ldots, p_n\}$, where $p_1 < p_2 < \cdots < p_n$. Then

   (c) G possesses a series of characteristic subgroups

$$G = G_0 \supset G_1 \supset \cdots \supset G_n = 1$$

   such that G_{i-1}/G_i is isomorphic to a Sylow p_i-subgroup of G for $i = 1, 2, \ldots, n$.

Proof. Let $F = F(G)$.

   (a) By Corollary 4.19, G' centralizes every chief factor U/V of G for which $U \subseteq F$. By Proposition 1.2, $G' \subseteq F$.

   (b) By (a), G/F is abelian. Hence $FS \lhd G$ and

$$G = FSN_G(S) = FN_G(S).$$

Clearly $T \lhd N_G(S)$. By (a), $T \subseteq F = (F \cap S) \times \mathcal{O}_{p'}(F)$. Therefore T is centralized by $\mathcal{O}_{p'}(F)$ and normalized by F. Thus T is normalized by G.

   (c) We use induction on $|G|$. Take $H \subseteq G$ such that

$$F \subseteq H \text{ and } H/F = \mathcal{O}_{p_1'}(G/F).$$

Then F contains a Sylow p_1-subgroup of H. By (a), G/H is a is a p_1-group. By Theorem 4.18(b), H has a normal p_1-complement K, which is then a normal p_1-complement of G. Since $F(K) \subseteq F(G)$, the result follows by induction. $\square$

5. Narrow p-Groups

In Section 1 we defined a p-group R to be *narrow* if $r(R) \leq 2$ or if R contains a subgroup R_0 of order p such that $C_R(R_0) = R_0 \times R_1$ for some cyclic subgroup R_1 of R. Narrow Sylow subgroups play an important role in the proof of the Feit-Thompson Theorem. In this section we will be concerned mainly with narrow p-groups that have rank greater than two. We show that these groups are *almost* as well behaved as p-groups of smaller rank.

Remark. For every odd prime p, there exists a narrow p-group R for which $r(R) = p \geq 3$, namely the wreath product $Z_p \wr Z_p$. These are discussed in [17, p. 324] and in [19, p. 82].

Lemma 5.1. Suppose p is an odd prime, R is a p-group, and $r(R) \geq 3$. Then

(a) $\mathrm{SCN}_3(R)$ is not empty, and
(b) if $E \in \mathcal{E}^2(R)$ and $E \lhd R$, then E is contained in an element of $\mathrm{SCN}_3(R)$.

Proof. Part (a) follows from Lemma 4.7. Take E as in (b). By (a) and Lemma 1.22, there exists a normal elementary abelian subgroup B of order p^3 in R. Let $B^* = EC_B(E)$. Then B^* is elementary abelian and $B^* \lhd R$. If $|B^*| \geq p^3$, then, by Proposition 4.4, B^* is contained in an element of $\mathrm{SCN}_3(R)$ and we are done.

Assume that $|B^*| < p^3$. Then $B^* = E$ and $E \supseteq C_B(E)$. Since $|E| = p^2$,

$$|B/C_B(E)| \leq p, \ |C_B(E)| \geq p^2, \text{ and } E = C_B(E) \subset B.$$

But then, since B is abelian, $B = C_B(E)$. This contradicts the previous sentence and completes the proof of Lemma 5.1. $\square$

Lemma 5.2. Suppose p is an odd prime, R is a p-group, $r(R) \geq 3$, and $E \in \mathcal{E}^2(R)$. Let $T = C_R(\Omega_1(Z_2(R)))$. Assume further that $E \in \mathcal{E}^*(R)$, that is, E is contained in no larger elementary abelian subgroup of R. Then

(a) E is not contained in T,
(b) $|\Omega_1(Z(R))| = p$ and $\Omega_1(Z_2(R)) \in \mathcal{E}^2(R)$, and
(c) T is a characteristic subgroup of index p in R.

Proof. Let $Z = \Omega_1(Z(R))$ and $W = \Omega_1(Z_2(R))$. Since $E \in \mathcal{E}^*(R)$ and EZ is elementary abelian, we have $EZ = E$. Thus $Z \subseteq E$. A similar argument shows that $r(C_R(E)) = 2$. Therefore

(5.1) $Z \subset E$ and $|Z| = p$.

By Lemma 4.5(c), W is a noncyclic group of exponent p. Consequently

(5.2) $Z \subset W$ and $[W, R] = [\Omega_1(Z_2(R)), R] \cap W \subseteq \Omega_1(Z(R)) = Z$.

Hence, by (5.1), $[E, W] \subseteq [R, W] \subseteq Z \subset E$. It follows that W normalizes E. Since $C_W(E)$ has exponent p and $E \in \mathcal{E}^*(R)$, we have $C_W(E) \subseteq E$. In fact,

$$Z \subseteq C_W(E) \subseteq E.$$

If $C_W(E) = E$, then $[E, R] \subset E$ by (5.2) and (5.1). But then $E \lhd R$, and hence $E \notin \mathcal{E}^*(R)$ by Lemma 5.1. Thus $C_W(E) \neq E$, so $C_W(E) = Z$. By (5.2),

$$1 < |W/Z| = |W/C_W(E)|.$$

As $|E| = p^2$ and $W/C_W(E)$ is isomorphic to a p-subgroup of $\mathrm{Aut}\, E$,

$$|W/Z| = p \text{ and } |W| = p^2.$$

We now have (a) and (b). Repeating our last argument, we see that

$$|R/T| = |R/C_R(W)| = p.$$

Clearly, T char R. This proves (c) and completes the proof of the lemma. □

Theorem 5.3. Suppose p is an odd prime, R is a p-group, and $r(R) \geq 3$. Then R is narrow if and only if $\mathcal{E}^2(R) \cap \mathcal{E}^*(R)$ is not empty (i.e., some elementary abelian subgroup of order p^2 in R is contained in no elementary abelian subgroup of order p^3 in R).

Suppose that R is narrow. Let $T = C_R(\Omega_1(Z_2(R)))$. Then

(a) no element of $\mathcal{E}^2(R) \cap \mathcal{E}^*(R)$ is contained in T,
(b) $|\Omega_1(Z(R))| = p$ and $\Omega_1(Z_2(R)) \in \mathcal{E}^2(R)$,
(c) T is a characteristic subgroup of index p in R, and
(d) if S is a subgroup of order p in R and $r(C_R(S)) \leq 2$, then $C_T(S)$ is cyclic, $S \cap R' = S \cap T = 1$, and $C_R(S) = S \times C_T(S)$.

Proof. Let $Z = \Omega_1(Z(R))$ and $T = C_R(\Omega_1(Z_2(R)))$.

First assume that R is narrow. Take a subgroup R_0 of order p such that $C_R(R_0) = R_0 \times R_1$ for some cyclic group R_1. Since

$$r(C_R(R_0)) \leq 2 < 3 \leq r(R),$$

$R_0 \not\subseteq Z$, and so $R_0 \cap Z = 1$. Hence $R_0 \subset R_0 \times Z \subseteq C_R(R_0) = R_0 \times R_1$. Thus $R_1 \neq 1$. Let

$$E = \Omega_1(C_R(R_0)) = R_0 \times \Omega_1(R_1).$$

Clearly $E \in \mathcal{E}^2(R) \cap \mathcal{E}^*(R)$. Thus $\mathcal{E}^2(R) \cap \mathcal{E}^*(R)$ is not empty. By Lemma 5.2, we obtain (a), (b), and (c).

Continuing to assume that R is narrow, we take any subgroup S of order p in R such that $r(C_R(S)) \leq 2$. The previous argument shows that $S \cap Z = 1$ and hence that $SZ \in \mathcal{E}^2(R)$. Since

(5.3) $r(SZ) \leq r(C_R(S)) \leq 2,$

$SZ \in \mathcal{E}^*(R)$. By (a), $SZ \not\subseteq T$. As $Z \subseteq T$ and $R' \subseteq T$, we have $S \cap T = 1$ and $S \cap R' = 1$. Consequently

$$R = ST \text{ and } C_R(S) = SC_T(S) = S \times C_T(S).$$

By (5.3), $r(C_T(S)) = 1$ and, by Lemma 4.5, $C_T(S)$ is cyclic. Thus we obtain (d).

Now we assume that $\mathcal{E}^2(R) \cap \mathcal{E}^*(R)$ is not empty and show that R is narrow. Take $E \in \mathcal{E}^2(R) \cap \mathcal{E}^*(R)$. Clearly $Z \subset E$. Hence $E = Z \times S$ for some subgroup S of order p in E. Since $E \in \mathcal{E}^*(R)$,

$$2 = r(C_R(E)) = r(C_R(S)).$$

Now, by the argument of the previous paragraph, $C_R(S) = S \times C_T(S)$ and $C_T(S)$ is cyclic. Therefore R is narrow. □

Corollary 5.4. Suppose p is an odd prime, R is a p-group, and $\mathrm{r}(R) \geq 3$. Then R is narrow if and only if $\mathrm{r}(C_R(S)) \leq 2$ for some subgroup S of order p in R.

Proof. If R is narrow, take R_0 as in the definition of a narrow p-group. Then $\mathrm{r}(C_R(R_0)) \leq 2$.

Conversely, suppose S is a subgroup of order p in R and $\mathrm{r}(C_R(S)) \leq 2$. Let $Z = \Omega_1(Z(R))$. Then $S \not\subseteq Z$ and hence $\mathrm{r}(SZ) > \mathrm{r}(S) = 1$. It follows that $SZ \in \mathcal{E}^2((R) \cap \mathcal{E}^*(R)$. Therefore, by Theorem 5.3, R is narrow. $\square$

Theorem 5.5. Suppose that p is an odd prime, R is a narrow p-group, and A is a solvable subgroup of $\mathrm{Aut}\, R$ having odd order. Then A has the following properties.

    (a) The factor group $A/\mathcal{O}_p(A)$ is an abelian p'-group.
    (b) If $\mathrm{r}(R) \geq 3$, then the order of every p'-element of A divides $p - 1$.
    (c) If $|A|$ is a prime that does not divide $p(p - 1)$ then $|A|$ divides $\frac{1}{2}(p + 1)$. If, in addition, $R = [R, A]$ and R is not abelian, then $|R| = p^3$.

Proof. By Theorem 1.13, R has a characteristic subgroup H of class at most two and of exponent p such that

$$(5.4) \qquad\qquad [R, H] \subseteq Z(H)$$

and $C_A(H)$ is a p-group. In particular, since $C_A(H) \lhd A$, it follows that $C_A(H) \subseteq \mathcal{O}_p(A)$. Therefore we can assume that $H \neq 1$.

We consider first the case in which $\mathrm{r}(R) \geq 3$. Take cyclic subgroups R_0 and R_1 such that $|R_0| = p$ and $C_R(R_0) = R_0 \times R_1$, and let $U = R_0 Z(H)$. If $R_0 \subseteq H$, then U is elementary abelian because H has exponent p, and $U \lhd H$ because of (5.4). Thus $\mathrm{m}(U) \leq \mathrm{r}(C_R(R_0)) \leq 2$. But this implies that either

$$\mathrm{m}(U) = 1,\ R_0 = U \lhd H,\ \text{and hence } R_0 \subseteq Z(R),\ \text{or}$$
$$\mathrm{m}(U) = 2 \text{ and, by Lemma 5.1, } \mathrm{r}(C_R(R_0)) \geq \mathrm{r}(C_R(U)) \geq 3.$$

Both are impossible. Therefore $R_0 \not\subseteq H$.

Now

$$C_H(R_0) \subseteq \Omega_1(C_R(R_0)) = R_0 \times \Omega_1(R_1).$$

Since $H \lhd R$, we have $H \cap Z(R) \neq 1$. Consequently

$$(5.5) \qquad\qquad |C_H(R_0)| = p.$$

For each nonnegative integer i, define inductively subgroups H_i of R by

$$H_0 = H \text{ and}$$
$$H_i = [R, H_{i-1}], \text{ for } i > 0.$$

Then, for each i, H_i char R and hence $H_i \lhd R$. Let $v \in R_0^{\#}$ and suppose that $p^n = |H|$. A short argument using the mapping $H \to [R, H]$ given by $x \mapsto [v, x] = v^{-1}x^{-1}vx$ shows that, for each i,

$$|H_{i+1}| \geq |[R_0, H_i]| \geq |H_i : C_{H_i}(v)| \geq p^{-1}|H_i|.$$

On the other hand, since R is nilpotent, $H_i = 1$ or $H_{i+1} \subset H_i$, for each i. Therefore we obtain an A-invariant chain

$$H = H_0 \supset H_1 \supset \cdots \supset H_n = 1$$

in which each factor group H_i/H_{i+1} has order p. Clearly the chain is stabilized (in the sense of Lemma 1.9) by A' and by α^{p-1} for each $\alpha \in A$. Recall that $C_A(H)$ is a p-group. Thus, by Lemma 1.9, we obtain (a), (b), and (c) in the case when $r(R) \geq 3$. (The hypotheses of (c) cannot occur in this case.)

We are left with the case in which $r(R) \leq 2$. Here (b) does not occur, and (a) follows from Lemma 4.17. To complete the proof, assume that $|A|$ is a prime that does not divide $p(p-1)$. Let $q = |A|$.

By Lemma 4.7, $\mathrm{SCN}_3(R)$ is empty and, since q does not divide $p - 1$, Lemma 4.14 implies that q divides $\frac{1}{2}(p+1)$.

Now suppose that $R = [R, A]$ and R is not abelian. By Theorem 4.16,

$$|\Omega_1(R)| = p^3 \text{ and } R/\Omega_1(R) \text{ is cyclic.}$$

As q does not divide $p - 1$, by **G**, Theorem 5.4.1, p. 189, A centralizes $R/\Omega_1(R)$. Since $[R, A] = R$, it follows that $R = \Omega_1(R)$. Thus $|R| = p^3$, which completes the proof of (c) and of the theorem. $\square$

Recall that a group G has p-length one if $G = \mathcal{O}_{p',p,p'}(G)$.

Theorem 5.6. Suppose G is a solvable group of odd order, $p \in \pi(G)$, and S is a narrow Sylow p-subgroup of G. If $r(S) \geq 3$, assume as well that G has p-length one. Then:

(a) p is the largest prime divisor of $|G/\mathcal{O}_{p'}(G)|$;
(b) if $p = 3$ or p is the smallest prime divisor of $|G|$, then G has a normal p-complement;
(c) G' has a normal p-complement;
(d) every p'-subgroup of G' is contained in $\mathcal{O}_{p'}(G')$; and
(e) $G/\mathcal{O}_{p',p}(G)$ is an abelian p'-group.

Proof. If $r(S) \leq 2$, use Theorem 4.18. If $r(S) \geq 3$, use Theorem 5.5 and the method of proof of Theorem 4.18. $\square$

Theorem 5.7. Suppose G is a solvable group of odd order, $p \in \pi(G)$, and E is an elementary abelian p-subgroup of $F(G)$. Assume as well that $r(C_{F(G)}(E)) \leq 2$. Then $G' \subseteq F(G)$.

Proof. By Proposition 1.2, it suffices to show that G' centralizes every chief factor U/V of G for which $U \subseteq F(G)$. Take such a chief factor U/V. Then U/V is a q-group for some prime q. We may assume that $U \subseteq \mathcal{O}_q(G)$.

Let $R = \mathcal{O}_q(G)$, $Z = \Omega_1(Z(R))$, and $C = C_R(E)$. Then $\mathrm{r}(C) \leq 2$. We claim that R is narrow. To prove this, we can assume that $\mathrm{r}(R) \geq 3$. Then $R \not\subseteq C$. Hence $q = p$ and $E \not\subseteq Z$. Therefore

$$1 \subset Z \subset EZ \text{ and } \mathrm{r}(C_R(EZ)) = \mathrm{r}(C) \leq 2.$$

It follows that $\mathrm{m}(EZ) = 2$ and $EZ \in \mathcal{E}^2(R) \cap \mathcal{E}^*(R)$. By Theorem 5.3, R is narrow.

Let $C_1 = C_G(U/V)$. Since R is narrow, Theorem 5.5 implies that G' induces a q-group of automorphisms on R by conjugation. Hence G' induces a q-group of automorphisms on U/V, that is, $G'C_1/C_1$ is a q-group. However, $\mathcal{O}_q(G/C_1) = 1$ because G acts irreducibly on U/V. Therefore $G' \subseteq C_1$, as desired. $\square$

6. Additional Results

Theorem 6.1 (P. Hall & G. Higman). Suppose G is a solvable group of odd order, p is a prime, and S is a Sylow p-subgroup of G. Then $\mathcal{O}_{p',p}(G)$ contains every abelian normal subgroup of S.

Proof. G, Theorem 6.5.2, p. 234. $\square$

Theorem 6.2. Suppose that G is a solvable group of odd order, p is a prime, and S is a Sylow p-subgroup of G. Then $Z(J(S))\mathcal{O}_{p'}(G) \lhd G$.

Proof. G, Theorem 6.5.1, p. 234 and Theorem 8.2.11, p. 279. $\square$

Remark. A substitute for this result (Theorem B.4) is proved in Appendix B, using the characteristic subgroup $L(S)$ of Puig instead of $J(S)$. Note that for $S \neq 1$, one has $L(S) \neq 1$ (Lemma B.1(f)).

Lemma 6.3. Let G be a solvable group.

(a) Suppose that H is a normal Hall subgroup of G and K is a complement of H in G. Assume that $H \subseteq G'$. Then $H = [H, K]$ and $C_H(K) \subseteq H'$.

(b) Suppose that G' is nilpotent and $|G/G'|$ is prime. Then G' is a Hall subgroup of G and $G' = [G, K]$ for every complement K of G' in G.

Proof. (a) Let $H^* = [H, K]$. Then $H^* \lhd HK = G$. Let $\overline{G} = G/H^*$, $\overline{H} = H/H^*$, and $\overline{K} = KH/H^*$. Then $\overline{G} = \overline{H} \times \overline{K}$ and $\overline{H} \subseteq \overline{G}' = \overline{H}' \times \overline{K}'$.

Hence $\overline{H} = \overline{H}'$. Since G is solvable, we have $\overline{H} = 1$, that is, $H = H^*$. Therefore, by Proposition 1.6(d) and Proposition 1.5(d),

$$H/H' = [H/H', KH'/H'] \times C_{H/H'}(KH'/H')$$
$$= [H, K]H'/H' \times C_H(K)H'/H'$$
$$= H/H' \times C_H(K)H'/H'.$$

This shows that $C_H(K)H'/H' = 1$. Thus $C_H(K) \subseteq H'$, as desired.

(b) Let $p = |G/G'|$. Then $G/\mathcal{O}_{p'}(G')$ is a p-group whose derived group has index p. Hence $G/\mathcal{O}_{p'}(G')$ is a cyclic group of order p and $G' = \mathcal{O}_{p'}(G')$. The rest of (b) follows from (a). $\square$

Theorem 6.4. Suppose G is a group, π is a set of primes, H is a π'-subgroup of G, and G_0 is a normal Hall subgroup of G. Assume that $G_0/F(G_0)$ and $(G/G_0)/F(G/G_0)$ are nilpotent. Assume further that H normalizes two π-subgroups J_1 and J_2 of G.

Then there exists an element $x \in \langle J_1, J_2 \rangle$ such that $\langle J_1^x, J_2 \rangle$ is a π-group and x centralizes H.

Proof. We use induction on $|G| + |H|$.

We can assume that $G \neq 1$. Let $M = G_0$ if $G_0 \neq 1$ and $M = G$ otherwise. Let $L = \langle J_1, J_2 \rangle$. Then

(6.1) M is a nonidentity normal Hall subgroup of G and
              $M/F(M)$ is nilpotent.

Since H normalizes J_1 and J_2, H normalizes L. We can assume that $G = LH$. Then G/L is a π'-group, and

(6.2) L contains every π-subgroup of G.

Suppose $\pi(F(G)) \nsubseteq \pi(H)$. Take $p \in \pi(F(G))$ such that $p \notin \pi(H)$. Let N be a minimal normal subgroup of G contained in $\mathcal{O}_p(F(G))$. Since $G/L = LH/L \cong H/(H \cap L)$, we have $N \subseteq L$. By induction, there exists $y \in L$ such that

(6.3) $\langle J_1^y, J_2 \rangle N/N$ is a π-group and y centralizes HN/N.

In this case, H is a Hall p'-subgroup of HN. Take $z \in N$ such that $(H^y)^z = H$. Let $L^* = \langle J_1^y, J_2 \rangle N$. Then $yz \in L$ and

$$[H, yz] \subseteq H \cap L = 1.$$

Thus yz centralizes H. Hence H normalizes J_1^{yz} and L^*. By (6.3), L^* has order relatively prime to the order of H. By Proposition 1.5, there exists $w \in C_{L^*}(H)$ such that $\langle J_1^{yzw}, J_2 \rangle$ is a π-group. Since

$$yzw \in \langle L, N, L^* \rangle = L \text{ and } yzw = (yz)w \in C_G(H),$$

we obtain the conclusion in this case.

Now assume that $\pi(F(G)) \subseteq \pi(H)$. Then

(6.4) $\mathcal{O}_\pi(G) = 1$ and $\pi(F(M)) \subseteq \pi(H)$.

By (6.1), M is a nonidentity normal Hall subgroup of G. Therefore M contains a Hall $\pi(F(M))$-subgroup of H that is not trivial. Let $B = H \cap M$ and let H^* be a complement of B in H. Then $B \neq 1$ and $|H^*| < |H|$. By induction, there exists an element $y \in L$ such that

$$\langle J_1{}^y, J_2 \rangle \text{ is a } \pi\text{-group and } y \text{ centralizes } H^*.$$

Let $K_1 = [J_1, B]$ and $F = F(M)$. Since $K_1 \subseteq J_1$,

(6.5) K_1 is a π-group.

By (6.1), M/F is nilpotent. Since B is a π'-group,

$$BF/F \subseteq \mathcal{O}_{\pi'}(M/F) \text{ and } K_1F/F = [BF/F, J_1F/F] \subseteq \mathcal{O}_{\pi'}(M/F).$$

By (6.5), K_1F/F is a π-group. Therefore $K_1F/F = 1$ and $K_1 \subseteq F$. Since F is nilpotent, (6.4) yields

$$K_1 \subseteq \mathcal{O}_\pi(F) \subseteq \mathcal{O}_\pi(G) = 1.$$

Thus B centralizes J_1. By symmetry, B centralizes J_2. Hence B centralizes L and y. Finally, since y centralizes H^* and $H = H^*B$, we see that y centralizes H.   □

Remark. If one weakens the hypothesis of Theorem 6.4 slightly, the conclusion need not hold. For example, one can have $\pi = \{3\}$, $G \cong \mathbf{GL}(2,3)$, $H \cong Z_2 \times Z_2$, $J_1 \cong J_2 \cong Z_3$, and $J_2 = J_1^y$ for some $y \in N_G(H) - H$.

Lemma 6.5. Suppose K, U, and H are subgroups of a solvable group G and

$$K \lhd G, \ G = KU, \ H \subseteq U, \text{ and } |H| \text{ is relatively prime to } |K|.$$

Then
  (a) $H \cap G' = H \cap U'$,
  (b) $N_G(H) = C_K(H)N_U(H)$, and
  (c) if $g \in G$ and $H^g \subseteq U$, then $g = cu$ for some $c \in C_K(H)$ and $u \in U$.

Proof. (a) Let $\pi = \pi(H)$. Since $K \lhd G = KU$, we know that $KU' \lhd G$ and G/KU' is abelian. Therefore

$$G' \subseteq KU' \text{ and } H \cap G' \subseteq U \cap G' \subseteq U \cap KU' = (U \cap K)U'.$$

Hence
$$(U \cap G')/U' \subseteq (U \cap K)U'/U' \cong (U \cap K)/(U' \cap K),$$

which is a π'-group. As $(H \cap G')U'/U'$ is a π-subgroup of $(U \cap G')/U'$, it follows that $(H \cap G')U' = U'$ and thus

$$H \cap G' \subseteq U'.$$

Consequently $H \cap G' \subseteq H \cap U'$. Since the opposite containment is obvious, $H \cap G' = H \cap U'$.

(b), (c) It is easy to derive (b) from (c), so we will prove only (c). Suppose $g \in G$ and $H^g \subseteq U$. Take $k \in K$ and $v \in U$ such that $kv = g$. Then

$$k^{-1}Hk = H^k = H^{gv^{-1}} \subseteq U^{v^{-1}} = U.$$

Now $k^{-1}Hk \subseteq HK$. Therefore $H^k \subseteq HK \cap U$. Clearly H is a Hall π-subgroup of HK and hence of $HK \cap U$. Consequently H and H^k are Hall π-subgroups of $HK \cap U$. Thus there exists an element $w \in HK \cap U$ such that $H^w = H^k$. Since

$$w \in HK \cap U = H(K \cap U),$$

we can assume that $w \in K \cap U$.

Let $c = kw^{-1}$ and $u = wv$. Then $c \in K$, $u \in U$, $g = kv = cu$, and

$$H^c = H^{kw^{-1}} = (H^w)^{w^{-1}} = H.$$

For each $h \in H$ we know that $h^c \in H$ and

$$h^c \equiv c^{-1}hc \equiv h \pmod{K}.$$

Therefore $h^c h^{-1} \in H \cap K = 1$, and hence c centralizes h. Thus $c \in C_K(H)$. This completes the proof of (c) and of the lemma. $\square$

Lemma 6.6. Suppose that G is a solvable group, p is a prime, and S is a Sylow p-subgroup of G. Assume that G has p-length one. Then:

(a) $S \subseteq SO_{p'}(G) = O_{p',p}(G)$ and $G = O_{p'}(G)N_G(S)$;
(b) if $S \subseteq G'$, then $S \subseteq (N_G(S))'$;
(c) if Y is a nonempty subset of S and $x \in G$ satisfies $Y^x \subseteq S$, then there exist $c \in C_G(Y)$ and $g \in N_G(S)$ such that $cg = x$; and
(d) if Q is a p-subgroup of G, then there exists $x \in C_G(Q \cap S)$ such that $Q^x \subseteq S$.

Proof. Let $M = O_{p'}(G)$ and $U = N_G(S)$.

(a) Since G has p-length one, $G = O_{p',p,p'}(G)$. Therefore $S \subseteq O_{p',p}(G)$. It follows that S is a Sylow p-subgroup of $O_{p',p}(G)$ and $O_{p',p}(G) = MS$. By the Frattini argument,

$$G = O_{p',p}(G)N_G(S) = MSU = MU = O_{p'}(G)N_G(S).$$

(b) Apply Lemma 6.5(a) with $H = S$ and $K = O_{p'}(G)$.

(c) Since $\langle Y \rangle$ and $\langle Y^x \rangle$ are contained in S, we can apply Lemma 6.5(c) with $K = M$ and $H = \langle Y \rangle$.

(d) By (a) and its proof, $Q \subseteq O_{p',p}(G) = MS$, and clearly S is a Sylow p-subgroup of MS. Hence there exist elements $x \in M$ and $y \in S$ such that $Q^{xy} \subseteq S$. Clearly $Q^x \subseteq S$. For $z \in Q \cap S$, we have

$$z^x z^{-1} \in S \text{ and } z^x \equiv x^{-1}zx \equiv z \pmod{M}.$$

Hence $z^x z^{-1} \in S \cap M = 1$. Thus $x \in C_G(Q \cap S)$. This completes the proof of (d) and of the lemma. $\square$

Theorem 6.7. Suppose that G is a solvable group, p is an odd prime, $E \in \mathcal{E}_p^*(G)$, and L is a p'-subgroup of G normalized by E. Assume that G has p-length one. Then $L \subseteq \mathcal{O}_{p'}(G)$.

Remark. The assumption of p-length one is unnecessary by a theorem of J. G. Thompson [18, Theorem X.1.12, p. 9]. However, the assumption that p is odd is necessary. For an example, let $G = \mathbf{SL}(2,3)$, $E = Z(G)$, and let L be any subgroup of order three in G.

Proof. Set $K = \mathcal{O}_{p'}(G)$ and let S be a Sylow p-subgroup of G that contains E. Then $E \in \mathcal{E}^*(S)$. Thus

$$(6.6) \qquad E = \Omega_1(C_S(E)) \text{ and } EK/K \in \mathcal{E}^*(SK/K).$$

Hence

$$EK/K \in \mathcal{E}_p^*(G/K).$$

Now if $LK/K \subseteq \mathcal{O}_{p'}(G/K) = 1$, then $L \subseteq K$. Therefore it suffices to prove the result for G/K, so we can assume that

$$K = \mathcal{O}_{p'}(G) = 1.$$

By Lemma 6.6, $\mathcal{O}_{p',p}(G) = KS = S$, so $S \lhd G$. Now L acts on S by conjugation, and L centralizes E because $[L, E] \subseteq L \cap S = 1$. Hence, by (6.6) and Corollary 1.12, L centralizes S.

Obviously S is a Sylow p-subgroup of $\mathcal{O}_{p',p}(G)$. Therefore, by Proposition 1.15(a),

$$L \subseteq C_G(S) \subseteq \mathcal{O}_{p',p}(G) = S.$$

Since L is a p'-group, $L = 1 \subseteq \mathcal{O}_{p'}(G)$, as desired. $\square$

CHAPTER II

The Uniqueness Theorem

In this chapter we introduce the minimal counterexample and begin to study it. As in the proof of the Feit-Hall-Thompson CN-group theorem, the major results concern the maximal subgroups of the minimal counterexample–their *structure* and the *relationships* between them. The first major result is known as the Uniqueness Theorem (Theorem 9.6).

Midway through the proof of the CN-group theorem, (e.g., after **G**, Theorem 14.2.3, p. 406), one can easily show that a subgroup of the form $Z_p \times Z_p$ is contained in a *unique* maximal subgroup of the minimal counterexample. In our case, the Uniqueness Theorem includes an analogous result for a subgroup of the form $Z_p \times Z_p \times Z_p$. It depends upon a deep preliminary result, the Thompson Transitivity Theorem (Theorem 7.6).

The other main ideas of the proof come from a paper of the first author [1], which makes essential use of the second author's ZJ-Theorem [11]. However, Thompson's Factorization Theorem [27], involving his J-subgroup and a variation thereof, already had strong applications to local analysis in **FT**.

7. The Transitivity Theorem

We now assume that the main theorem is false. Henceforth in these notes we let G denote a fixed counterexample of minimal order. Of course G is a nonabelian simple group. We also fix the following notation:

$$\mathscr{M} \ = \ \text{the set of all maximal subgroups of } G,$$

$\mathscr{M}(H) \ = \ $ the set of all maximal subgroups of G that contain H (for each proper subgroup H of G),

$\mathscr{U} \ = \ $ the set of all proper subgroups H of G for which $\mathscr{M}(H)$ has a unique element.

(The set $\mathscr{U}$ is not related to the sets $\mathscr{U}(P)$ and $\mathscr{U}(p)$ defined in **FT**.)

For each prime p, recall that p' is the set of all primes other than p. Let $\mathrm{SCN}_3(p)$ be the set of all subgroups A of G for which $A \in \mathrm{SCN}_3(P)$ for some Sylow p-subgroup P of G.

For a set of primes π, recall that π' is the set of all primes not contained in π. We also introduce the following notation. Whenever A is a π'-subgroup of G and H is a subgroup of G, $\mathit{U}_H(A; \pi)$ denotes the set of all π-subgroups of H normalized by A and $\mathit{U}_H{}^*(A; \pi)$ denotes the set of all maximal elements of $\mathit{U}_H(A; \pi)$ under inclusion. If $\pi = \{q\}$ for some prime q, we also write $\mathit{U}_H(A; q)$ and $\mathit{U}_H{}^*(A; q)$ for $\mathit{U}_H(A; \pi)$ and $\mathit{U}_H{}^*(A; \pi)$, respectively.

Hypothesis 7.1.

(1) The group A is a nonidentity proper subgroup of G, $\pi = \pi(A)$, and $K = \mathcal{O}_{\pi'}(C_G(A))$.

(2) Whenever X is a proper subgroup of G that contains A, we have $\langle\, \mathit{U}_X(A; \pi') \,\rangle = \mathcal{O}_{\pi'}(X)$.

Note that in Hypothesis 7.1, K is the set of all π'-elements in $C_G(A)$. Clearly, for every $q \in \pi'$, K acts upon $\mathit{U}_G{}^*(A; q)$ by conjugation. In this section we will give some sufficient conditions for the action to be transitive.

Lemma 7.1. Assume Hypothesis 7.1. Suppose, for some prime $q \in \pi'$, that $Q_1, Q_2 \in \mathit{U}_G{}^*(A; q)$, and that there exists a proper subgroup H of G such that

$$A \subseteq H, \quad H \cap Q_1 \neq 1, \quad \text{and} \quad H \cap Q_2 \neq 1.$$

Then $Q_2 = Q_1{}^k$ for some $k \in K$.

Proof. We proceed by induction on $|G|_q / |Q_1 \cap Q_2|$. Since $A \subseteq H$, A normalizes $H \cap Q_1$ and $H \cap Q_2$. By Hypothesis 7.1, $\mathcal{O}_{\pi'}(H)$ contains $H \cap Q_1$ and $H \cap Q_2$. By Proposition 1.5, $H \cap Q_i$ is contained in an A-invariant Sylow q-subgroup R_i of $\mathcal{O}_{\pi'}(H)$ (for $i = 1, 2$), and $R_1{}^h = R_2$ for some $h \in C_G(A) \cap \mathcal{O}_{\pi'}(H)$. Since h is a π'-element of $C_G(A)$, we know $h \in K$. Take $Q_3 \in \mathit{U}_G{}^*(A; q)$ such that $R_2 \subseteq Q_3$. Then

$$Q_1{}^h \in \mathit{U}_G{}^*(A; q),$$

$$1 \subset Q_1{}^h \cap H \quad \text{and} \quad Q_2 \cap H \subseteq R_2 \subseteq Q_3.$$

Therefore

$$(7.1) \qquad 1 \subset Q_1{}^h \cap H \subseteq Q_1{}^h \cap Q_3 \quad \text{and} \quad 1 \subset Q_2 \cap H \subseteq Q_2 \cap Q_3.$$

Now suppose $Q_1 \cap Q_2 = 1$. By (7.1), $|Q_1 \cap Q_2| < |Q_1{}^h \cap Q_3|$. Hence, by induction, there exists $f \in K$ such that $(Q_1{}^h)^f = Q_3$. Similarly, $Q_3{}^g = Q_2$ for some $g \in K$. Since $h \in K$, we have

$$hfg \in K \quad \text{and} \quad Q_1{}^{hfg} = Q_3{}^g = Q_2.$$

Now suppose $Q_1 \cap Q_2 \neq 1$. Let $Q = Q_1 \cap Q_2$. Then our hypothesis is satisfied with $N_G(Q)$ in place of H, and we can assume that $H = N_G(Q)$.

The argument of the previous paragraph shows that we can assume further that $|Q| \geq |Q_1{}^h \cap Q_3|$ or $|Q| \geq |Q_2 \cap Q_3|$. Since (7.1) yields

$$|Q_1{}^h \cap Q_3| \geq |Q_1{}^h \cap H| = |Q_1 \cap H| = |N_{Q_1}(Q)|$$

and likewise $|Q_2 \cap Q_3| \geq |N_{Q_2}(Q)|$, it follows that $|Q| \geq |N_{Q_i}(Q)|$ for some $i = 1$ or 2. As Q_i is a p-group,

$$Q_i = Q = Q_1 \cap Q_2 \subseteq Q_{3-i}.$$

Since $Q_i \in \mathcal{U}_G{}^*(A; q)$, we have $Q_i = Q_{3-i}$. Thus $Q_2 = Q_1{}^k$ for $k = 1$. This completes the proof of Lemma 7.1. $\quad\square$

Theorem 7.2. Assume Hypothesis 7.1 and let $q \in \pi'$. Suppose $\mathrm{m}(Z(A)) \geq 3$. Then K acts transitively on $\mathcal{U}_G{}^*(A; q)$.

Proof. By hypothesis, $Z(A)$ contains an elementary abelian subgroup B of order p^3 for some prime p. Clearly $q \neq p$. Take $Q_1, Q_2 \in \mathcal{U}_G{}^*(A; q)$. We wish to prove that $Q_1{}^k = Q_2$ for some $k \in K$. If $Q_1 = 1$, then $Q_2 = 1$, and we can take $k = 1$. Assume that $Q_1 \neq 1$.

By Proposition 1.16,

$$Q_1 = \langle\, C_{Q_1}(C) \mid C \subseteq B \text{ and } B/C \text{ is cyclic} \,\rangle.$$

Thus $C_{Q_1}(C) \neq 1$ for some subgroup C of order p^2 in B. Similarly, since C is not cyclic, $C_{Q_2}(z) \neq 1$ for some $z \in C^\#$. As $C_{Q_1}(C) \subseteq C_G(z)$, the desired conclusion follows from Lemma 7.1 with $H = C_G(z)$. $\quad\square$

Theorem 7.3. Assume Hypothesis 7.1 and let $q \in \pi'$. Suppose $\mathrm{m}(Z(A)) \geq 2$ and $q \in \pi(C_G(A))$. Then K acts transitively on $\mathcal{U}_G{}^*(A; q)$.

Proof. Take $B \in \mathcal{E}_p{}^2(Z(A))$ and $Q_1, Q_2 \in \mathcal{U}_G{}^*(A; q)$. Let R be an element of $\mathcal{U}_G{}^*(A; q)$ that contains a Sylow q-subgroup of $C_G(A)$. Then $C_R(A) \neq 1$ and hence $Q_1, Q_2 \neq 1$.

By Proposition 1.16, $C_{Q_1}(x) \neq 1$ for some $x \in B^\#$. By Lemma 7.1 with $H = C_G(x)$, we have $Q_1{}^f = R$ for some $f \in K$. Similarly, $R^g = Q_2$ for some $g \in K$. Thus

$$fg \in K \text{ and } Q_1{}^{fg} = Q_2.$$

Since Q_1 and Q_2 are arbitrary elements of $\mathcal{U}_G{}^*(A; q)$, we are done. $\quad\square$

Theorem 7.4. Assume Hypothesis 7.1 and let $q \in \pi'$. Suppose that P is a proper π-subgroup of G that contains A as a subnormal subgroup and that K acts transitively on $\mathcal{U}_G{}^*(A; q)$. Then

    (a) $C_K(P) = \mathcal{O}_{\pi'}(C_G(P))$,
    (b) $\mathcal{O}_{\pi'}(C_G(P))$ acts transitively on $\mathcal{U}_G{}^*(P; q)$,
    (c) $\mathcal{U}_G{}^*(P; q) \subseteq \mathcal{U}_G{}^*(A; q)$, and
    (d) for every $Q \in \mathcal{U}_G{}^*(P; q)$ we have $P \cap N_G(P)' \subseteq N_G(Q)'$ and $N_G(P) = \mathcal{O}_{\pi'}(C_G(P))(N_G(P) \cap N_G(Q))$.

Proof. Recall that K equals $\mathcal{O}_{\pi'}(C_G(A))$ and contains every π'-element of $C_G(A)$. Since $A \subseteq P$, we have $C_G(P) \subseteq C_G(A)$. This proves (a).

To prove the other parts of the theorem we use induction on $|P : A|$. By the definition of a subnormal subgroup, there exists a normal series of P that contains A. We can refine such a series to a composition series

$$1 = P_0 \lhd P_1 \lhd \cdots \lhd P_{n-1} \lhd P_n = P,$$

where $P_k = A$ and $1 \leq k \leq n$.

Assume first that $k \leq n-2$. Let $B = P_{n-1}$. Clearly B is a π-group. Since $|B : A| < |P : A|$, by induction, $\mathcal{O}_{\pi'}(C_G(B))$ is transitive on $\mathcal{U}_G^*(B; q)$. Moreover, since $\mathcal{U}_X(B; \pi') \subseteq \mathcal{U}_X(A; \pi')$, Hypothesis 7.1 is satisfied with B and $\mathcal{O}_{\pi'}(C_G(B))$ in place of A and K. As $|P : B| < |P : A|$, (b), (c), and (d) follow by induction.

For the remainder of the proof we assume that $k > n-2$. Then $A = P_{n-1}$ or $A = P_n = P$. Thus $A \lhd P$, and

(7.2)   either $A = P$ or $|P/A|$ is a prime in π.

Let $\Omega = \mathcal{U}_G^*(A; q)$. Then P acts on Ω by conjugation and hence P/A acts on Ω. By hypothesis, the π'-group K acts transitively on Ω by conjugation. Therefore $|\Omega|$ divides $|K|$. Consequently, by (7.2), P/A fixes some element of Ω. Thus

(7.3) P normalizes some element of $\mathcal{U}_G^*(A; q)$.

Suppose first that $1 \in \mathcal{U}_G^*(P; q)$. Then $\{1\} = \mathcal{U}_G^*(P; q) = \mathcal{U}_G(P; q)$. Therefore (b) and (d) are trivial. Moreover, (7.3) shows that $1 \in \mathcal{U}_G^*(A; q)$ and thus yields (c).

Now assume that $1 \notin \mathcal{U}_G^*(P; q)$. To prove (c), take any $Q \in \mathcal{U}_G^*(P; q)$. Then A normalizes Q, so Q is contained in some element Q_1 of $\mathcal{U}_G^*(A; q)$. Now $N_G(Q)$ contains P and $N_{Q_1}(Q)$. By Hypothesis 7.1,

$$N_{Q_1}(Q) \subseteq \mathcal{O}_{\pi'}(N_G(Q)).$$

By Proposition 1.5, Q is contained in a P-invariant Sylow q-subgroup Q_2 of $\mathcal{O}_{\pi'}(N_G(Q))$. Since $Q \in \mathcal{U}_G^*(P; q)$, we know $Q = Q_2$. Therefore

$$|Q| = |Q_2| = |\mathcal{O}_{\pi'}(N_G(Q))|_q \geq |N_{Q_1}(Q)| \geq |Q|.$$

It follows that $Q = N_{Q_1}(Q)$ and hence that $Q = Q_1 \in \mathcal{U}_G^*(A; q)$. This proves (c).

To prove (b), let us take any $Q_1, Q_2 \in \mathcal{U}_G^*(P; q)$. First notice that $K = \mathcal{O}_{\pi'}(C_G(A)) \lhd N_G(A)$, so that KP is a group. By (c) and the hypotheses, $Q_2 = Q_1{}^k$ for some $k \in K$. Therefore

$$P \subseteq N_{KP}(Q_2) \text{ and } P^k \subseteq N_{KP}(Q_2).$$

Since PK is solvable, P is a π-group, and K is a π'-group, it follows that P and P^k are Hall π-subgroups of $N_{KP}(Q_2)$ and hence are conjugate in $N_{KP}(Q_2)$. As

$$N_{KP}(Q_2) = KP \cap N_G(Q_2) = (K \cap N_G(Q_2))P,$$

$(P^k)^g = P$ for some $g \in K \cap N_G(Q_2)$. Thus

$$Q_1{}^{kg} = Q_2{}^g = Q_2 \text{ and } kg \in N_K(P) = C_K(P).$$

This proves (b).

To prove (d), take any $Q \in \mathit{H}_G{}^*(P;q)$ and let $L = N_G(P) \cap N_G(Q)$. By (a) and (b), $N_G(P) = LC_K(P) = C_K(P)L$. Therefore, by Lemma 6.5,

$$P \cap (N_G(P))' = P \cap L' \subseteq L' \subseteq (N_G(Q))'.$$

This proves (d) and completes the proof of the theorem. $\square$

Proposition 7.5. Suppose $p \in \pi(G)$ and A is an abelian p-subgroup of G. Assume that either

(1) $A = \{ x \in C_G(A) \mid x^p = 1 \}$ and every proper subgroup of G has p-length one, or
(2) $A \in \mathrm{SCN}_2(P)$ for some Sylow p-subgroup P of G.

Then A satisfies Hypothesis 7.1.

Remark. For $A \in \mathrm{SCN}_2(P)$ this result is a special case of Theorem 8.5.1 of **G**, from which our proof is derived.

Proof. For case (1), Hypothesis 7.1 follows easily from Theorem 6.7. Therefore we will assume case (2) for the remainder of this proof.

To conform with the notation of Hypothesis 7.1, let $\pi = \pi(A) = \{p\}$ and $K = O_{p'}(C_G(A))$. In addition, let $Z = \Omega_1(Z(P))$.

Since $A \lhd P$, we have $\Omega_1(A) \lhd P$. Now $Z(P) \subseteq A$. If $Z(P)$ is not cyclic, take $B \in \mathcal{E}_p{}^2(Z)$. If $Z(P)$ is cyclic, then $|Z| = p$ and

$$(\Omega_1(A)/Z) \cap Z(P/Z) \neq 1,$$

by **G**, Theorem 2.6.4, p. 31. Thus the intersection contains a subgroup B/Z of order p. Either way

(7.4) $$B \in \mathcal{E}_p{}^2(A) \text{ and } B \lhd P.$$

Now suppose that X is an arbitrary proper subgroup of G that contains A and that $Y \in \mathit{H}_X(A; p')$. To prove our conclusion, it suffices to show that $Y \subseteq O_{p'}(X)$. We will verify two special cases and then the general case.

First suppose that $X = C_G(b)$ for some $b \in B^{\#} \cap Z$. Then P is a Sylow p-subgroup of X. Since A is an abelian normal subgroup of P, Theorem 6.1 yields

$$A \subseteq O_{p',p}(X).$$

Let $\overline{H} = H\mathcal{O}_{p'}(X)/\mathcal{O}_{p'}(X)$ for every subgroup H of X. Then

$$[\overline{A}, \overline{Y}] \subseteq \mathcal{O}_p(\overline{X}) \cap \overline{Y} = 1, \text{ and so,}$$
$$\overline{A} \subseteq C_{\mathcal{O}_p(\overline{X})}(\overline{Y}).$$

Since $C_P(A) \subseteq A$, we have $C_{\mathcal{O}_p(\overline{X})}(\overline{A}) \subseteq \overline{A}$. By Proposition 1.10 (with $\overline{Y}$ and $\mathcal{O}_p(\overline{X})$ in place of A and G), $\overline{Y}$ centralizes $\mathcal{O}_p(\overline{X})$. Next, by Proposition 1.15(a),

$$C_{\overline{X}}(\mathcal{O}_p(\overline{X})) \subseteq \mathcal{O}_{p',p}(\overline{X}) = \mathcal{O}_p(\overline{X}).$$

Consequently, as $\overline{Y}$ is a p'-group, $\overline{Y} = 1$. This shows that $Y \subseteq \mathcal{O}_{p'}(X)$.

Now suppose $X = C_G(b)$ for any $b \in B^\#$. By the paragraph above, we can assume that $|Z| = p$ and $B = \langle b \rangle \times Z$. Let $P_1 = C_P(b)$ and take a Sylow p-subgroup P_2 of X that contains P_1. By (7.4),

$$P/P_1 = P/C_P(B) \cong Z_p.$$

Therefore $|P_2 : P_1| \le p$ and $P_1 \lhd P_2$. Hence $Z \subseteq Z(P_1)$ and $Z(P_1) \lhd P_2$. By Theorem 6.1,

$$Z \subseteq Z(P_1) \subseteq \mathcal{O}_{p',p}(X).$$

Thus $[Y, Z] \subseteq Y \cap \mathcal{O}_{p',p}(X) \subseteq \mathcal{O}_{p'}(X)$. Since A normalizes $C_Y(Z)$, the previous paragraph shows that $C_Y(Z) \subseteq \mathcal{O}_{p'}(C_G(Z))$. Consequently we have $C_Y(Z) \subseteq \mathcal{O}_{p'}(C_X(Z))$ and, by Proposition 1.15(b), $C_Y(Z) \subseteq \mathcal{O}_{p'}(X)$. Thus

$$Y = C_Y(Z)[Y, Z] \subseteq \mathcal{O}_{p'}(X).$$

Assume once again that X is arbitrary. By Proposition 1.16,

(7.5) $$Y = \langle C_Y(b) \mid b \in B^\# \rangle.$$

For each $b \in B^\#$, A normalizes $C_Y(b)$ and the special cases above show that $C_Y(b) \subseteq \mathcal{O}_{p'}(C_G(b))$. Hence

$$C_Y(b) \subseteq \mathcal{O}_{p'}(C_X(b)) \subseteq \mathcal{O}_{p'}(X)$$

by Proposition 1.15(b). Finally, $Y \subseteq \mathcal{O}_{p'}(X)$ by (7.5), which completes the proof of the proposition. $\square$

The following result is a special case of the Thompson Transitivity Theorem (**G**, Theorem 8.5.4, p. 292).

Theorem 7.6 (Thompson Transitivity Theorem). Suppose that $p \in \pi(G)$, $A \in \mathrm{SCN}_3(p)$, and $q \in p'$. Then $\mathcal{O}_{p'}(C_G(A))$ acts transitively on $\mathit{M}_G^*(A; q)$ by conjugation.

Proof. This follows from Proposition 7.5 and Theorem 7.2. $\square$

8. The Fitting Subgroup of a Maximal Subgroup

Having proved the Transitivity Theorem, we now embark on the proof of the Uniqueness Theorem. In this section, we prove (Theorem 8.1) that if a maximal subgroup M is *large* in the sense that $r(F(M)) \geq 3$, then certain *large* subgroups of $F(M)$ lie in $\mathscr{U}$. Here and later we will frequently use the simplicity of G to assert that if L is a nonidentity normal subgroup of a maximal subgroup M, then $N_G(L) = M$.

Theorem 8.1. Suppose $M \in \mathscr{M}$, $p \in \pi(F(M))$, and $A_0 \in \mathscr{E}_p^*(F(M))$. Assume that $m(A_0) \geq 3$. Let P be a Sylow p-subgroup of M.

(a) If $F(M)$ is not a p-group, then $C_{F(M)}(A_0) \in \mathscr{U}$.

(b) If $F(M)$ is a p-group, then P is a Sylow p-subgroup of G and every element of $\mathrm{SCN}_3(P)$ is contained in $F(M)$ and belongs to $\mathscr{U}$.

Remark. Recall that, by Lemma 5.1, $\mathrm{SCN}_3(P)$ is not empty.

Proof. Let $F = F(M)$. For each nilpotent subgroup K of G and each prime q, let $K_q = \mathcal{O}_q(K)$. We handle parts (a) and (b) separately.

Proof of (a). Let $\pi = \pi(F)$ and $A = C_F(A_0)$. Then $\pi(A) = \pi$ because

(8.1) $$Z(F) \subseteq C_F(A_0) = A \subseteq F.$$

Note that, for every $q \in \pi$,

(8.2) $$C_G(A) \subseteq C_G(A_q) \subseteq C_G(Z(F)_q) \subseteq N_G(Z(F)_q) = M.$$

Suppose x is a π'-element of $C_G(A)$. Let $C = C_F(x) = C_F(\langle x \rangle)$. By (8.2), $x \in M$, so x operates on F by conjugation. Thus

$$C_F(C) \subseteq C_F(A) \subseteq C_F(A_0) = A \subseteq C.$$

By Propositions 1.10 and 1.3,

$$x \in C_M(F) = C_M(F(M)) \subseteq F.$$

Since x is a π'-element, $x = 1$. Thus, by (8.2),

(8.3) $$C_G(A) \text{ is a } \pi\text{-subgroup of } M.$$

Now we verify Hypothesis 7.1 for A. Take an arbitrary proper subgroup X of G that contains A and an A-invariant π'-subgroup Y of X. It will suffice to show that $Y \subseteq \mathcal{O}_{\pi'}(X)$.

Take any $q \in \pi$. By (8.2), $C_Y(A_q) \subseteq M$. Then, since Y is a π'-group,

$$[C_Y(A_q), A] \subseteq Y \cap [M, A] \subseteq Y \cap F(M) = 1.$$

Consequently $C_Y(A_q)$ centralizes A. By (8.3), $C_Y(A_q) = 1$. Thus, by Proposition 1.6(a),

(8.4) $$Y = C_Y(A_q)[Y, A_q] = [Y, A_q].$$

Since, by hypothesis, $|\pi| \geq 2$, there exists $r \neq q$ in π. Since, by (8.2), $N_G(Z(F)_q) = M$, we have

$$A_r \subseteq F_r \subseteq \mathcal{O}_{q'}(M) = \mathcal{O}_{q'}(N_G(Z(F)_q)).$$

Therefore $A_r \subseteq \mathcal{O}_{q'}(N_X(Z(F)_q))$. Moreover,

$$[\mathcal{O}_{q'}(N_X(Z(F)_q)), Z(F)_q] \subseteq \mathcal{O}_{q'}(N_X(Z(F)_q)) \cap Z(F)_q = 1,$$

so $\mathcal{O}_{q'}(N_X(Z(F)_q)) = \mathcal{O}_{q'}(C_X(Z(F)_q))$. Hence, by Proposition 1.15(b),

(8.5) $\qquad A_r \subseteq \mathcal{O}_{q'}(C_X(Z(F)_q)) \subseteq \mathcal{O}_{q'}(X)$ (for $q \neq r$ in π).

Consequently, by (8.4), with r in place of q,

$$Y = [Y, A_r] \subseteq \mathcal{O}_{q'}(X).$$

Since q was chosen arbitrarily in π,

$$Y \subseteq \bigcap_{q \in \pi} \mathcal{O}_{q'}(X) = \mathcal{O}_{\pi'}(X).$$

This proves Hypothesis 7.1 for A.

Now take q to be some prime in π'. Since $\mathrm{m}(Z(A)) \geq \mathrm{m}(A_0) \geq 3$, we can conclude from Theorem 7.2 that $\mathcal{O}_{\pi'}(C_G(A))$ acts transitively on $\mathit{H}_G^*(A; q)$ by conjugation. By (8.3), $\mathcal{O}_{\pi'}(C_G(A)) = 1$. Thus

$$\mathit{H}_G^*(A; q) = \{Q\} \text{ for some } q\text{-subgroup } Q \text{ of } G.$$

Since F is nilpotent, $A \lhd\lhd F$. By Theorem 7.4,

$$\mathit{H}_G^*(F; q) \subseteq \mathit{H}_G^*(A; q) = \{Q\}.$$

Therefore $\mathit{H}_G^*(F; q) = \{Q\}$ and M normalizes Q. As M is a maximal subgroup of G, $MQ = M$ and $Q \subseteq M$. Hence $Q \lhd M$ and $Q \subseteq F(M) = F$. Since $\pi = \pi(F)$ and $q \in \pi'$, we have $Q = 1$. Thus

(8.6) $\qquad \mathit{H}_G^*(A; q) = \{1\}$, for every $q \in \pi'$.

To prove that $A \in \mathscr{U}$, take $H \in \mathscr{M}(A)$. We will show that $H = M$.

Let $D = F(H)$ and $\sigma = \pi(D)$. By (8.6), we have $\sigma \subseteq \pi$. Since $F(\mathcal{O}_{\sigma'}(H)) \subseteq \mathcal{O}_{\sigma'}(F(H)) = 1$, we have $\mathcal{O}_{\sigma'}(H) = 1$. By (8.1) and (8.5) for $X = H$,

$$\mathcal{O}_{\sigma'}(Z(F)) \subseteq \mathcal{O}_{\sigma'}(A) = \langle A_r \mid r \in \pi \cap \sigma' \rangle \subseteq \bigcap_{q \in \sigma} \mathcal{O}_{q'}(X) = \mathcal{O}_{\sigma'}(H) = 1.$$

As F is nilpotent, $\mathcal{O}_{\sigma'}(F) = 1$, so $\pi = \pi(F) \subseteq \sigma$. Since $\sigma \subseteq \pi$, we have $\sigma = \pi$.

Recall that $|\pi| \geq 2$. For each $q \in \pi$,

(8.7) $\qquad [D_q, \mathcal{O}_{q'}(A)] \subseteq [\mathcal{O}_q(H), \mathcal{O}_{q'}(H)] = 1$, by (8.5) for $X = H$,

and $D_q \subseteq C_G(\mathcal{O}_{q'}(A)) \subseteq M$ by (8.2). Hence $D \subseteq M$.

By (8.7), A_p centralizes $\mathcal{O}_{p'}(D)$. Since

$$\mathcal{O}_{p'}(D) = \mathcal{O}_{p'}(F(H)) = F(\mathcal{O}_{p'}(H)),$$

Proposition 1.4 implies that A_p centralizes $\mathcal{O}_{p'}(H)$. By (8.2), $\mathcal{O}_{p'}(H) \subseteq M$. Since

$$D \subseteq M \text{ and } \mathcal{O}_{p'}(H) = \mathcal{O}_{p'}(N_G(D_p)),$$

we have $\mathcal{O}_{p'}(H) \subseteq \mathcal{O}_{p'}(N_M(D_p))$. Furthermore, since

$$[\mathcal{O}_{p'}(N_M(D_p)), D_p] \subseteq \mathcal{O}_{p'}(N_M(D_p)) \cap D_p = 1,$$

it follows that $\mathcal{O}_{p'}(N_M(D_p)) \subseteq \mathcal{O}_{p'}(C_M(D_p))$. Thus Proposition 1.15(b) yields

(8.8) $\mathcal{O}_{p'}(H) \subseteq \mathcal{O}_{p'}(M)$.

Since A_0 is a p-subgroup of F, we have $\mathcal{O}_{p'}(F) \subseteq C_F(A_0) = A$. Thus $\mathcal{O}_{p'}(F) = \mathcal{O}_{p'}(A)$. By (8.7), D_p centralizes $\mathcal{O}_{p'}(F)$. Since

$$D_p \subseteq D \subseteq M \text{ and } \mathcal{O}_{p'}(F) = \mathcal{O}_{p'}(F(M)) = F(\mathcal{O}_{p'}(M)),$$

Proposition 1.4 shows that D_p centralizes $\mathcal{O}_{p'}(M)$. Thus it follows that $\mathcal{O}_{p'}(M) \subseteq C_G(D_p) \subseteq H$.

By (8.2), $N_G(Z(F)_p) = M$. Therefore $\mathcal{O}_{p'}(M) \subseteq \mathcal{O}_{p'}(N_H(Z(F)_p))$. By (8.5), with $X = H$ and $q = p$, we have $\mathcal{O}_{p'}(M) \subseteq \mathcal{O}_{p'}(H)$. Hence, by (8.8),

$$\mathcal{O}_{p'}(H) = \mathcal{O}_{p'}(M) \text{ and } H = N_G(\mathcal{O}_{p'}(H)) = N_G(\mathcal{O}_{p'}(M)) = M,$$

which completes the proof of (a). □

Proof of (b). Take $A \in \mathrm{SCN}_3(P)$. Since $F(\mathcal{O}_{p'}(M)) \subseteq \mathcal{O}_{p'}(F) = 1$, we have

(8.9) $\mathcal{O}_{p'}(M) = 1$ and $F = \mathcal{O}_p(M) = \mathcal{O}_{p',p}(M)$.

By Theorem 6.1, $A \subseteq F$. Hence $Z(F) \subseteq \mathcal{O}_p(M) \cap C_G(A) \subseteq C_P(A) = A$. As $Z(F) \lhd M$,

(8.10) $C_G(A) \subseteq C_G(Z(F)) \subseteq N_G(Z(F)) = M$.

By (8.9) and Theorem 6.2, we know that $Z(J(P)) \lhd M$. Consequently $N_G(P) \subseteq N_G(Z(J(P))) = M$. Hence P is a Sylow p-subgroup of G and $A \in \mathrm{SCN}_3(p)$.

Let $A^* = \mathcal{O}_{p'}(C_G(A))$. By (8.10), $A^* \subseteq M$. Since

$$C_F(C_F(A^*)) \subseteq C_F(A) = A \subseteq C_F(A^*),$$

Propositions 1.10 and 1.3 imply that $A^* \subseteq C_M(F) \subseteq F$. As F is p-group,

(8.11) $1 = A^* = \mathcal{O}_{p'}(C_G(A))$.

Take any prime $q \in p'$. By (8.11) and the Thompson Transitivity Theorem (Theorem 7.6), $\mathcal{U}_G^*(A; q)$ contains a unique element, say, Q. Thus

$N_G(A)$ normalizes Q, and hence F normalizes Q. Since $A \subseteq F$ and $F \lhd M$, it follows that Q is the unique element of $\mathcal{U}_G^*(F; q)$ and that M normalizes Q. As M is a maximal subgroup of G,

$$MQ = M \text{ and } Q \subseteq \mathcal{O}_q(M) = 1.$$

Thus $\mathcal{U}_G^*(A; q) = \{1\} = \mathcal{U}_G(A; q)$. Take any $Y \in \mathcal{U}_G(A; p')$. For each $q \in p'$, we have $\mathcal{O}_q(Y) \in \mathcal{U}_G(A; q) = \{1\}$. Therefore $F(Y) = 1$. Since Y is solvable, $Y = 1$. Thus

$$(8.12) \qquad\qquad \mathcal{U}_G(A; p') = \{1\}.$$

To complete the proof of (b) we must show that $A \in \mathcal{U}$. Suppose that $A \notin \mathcal{U}$. Take $H \in \mathcal{M}(A)$ such that $|H \cap M|_p$ is maximal subject to $H \neq M$. Let R be a Sylow p-subgroup of $H \cap M$ containing A. If $|R| < |P|$, then

$$|H \cap M|_p = |R| < |N_M(R)|_p = |N_G(R) \cap M|_p$$

and $\{M\} = \mathcal{M}(N_G(R))$ by our choice of H. Therefore we know that $|N_H(R)|_p = |H \cap N_G(R)|_p \leq |H \cap M|_p = |R|$, and so R is a Sylow p-subgroup of H regardless of whether $|R| < |P|$ or $|R| = |P|$. By (8.12) and Theorem 6.2,

$$(8.13) \qquad\qquad \mathcal{O}_{p'}(H) = 1 \text{ and } Z(J(R)) \lhd H.$$

Thus $N_G(R) \subseteq N_G(Z(J(R))) = H$, so R is a Sylow p-subgroup of G and of M. By (8.9), (8.13), and Theorem 6.2, $M = N_G(Z(J(R))) = H$. This contradicts our choice of H and completes the proof of Theorem 8.1. $\square$

Remark. The theorem above has slightly different conclusions, but very different arguments according to whether (a) $F(M)$ is not a p-group or (b) $F(M)$ is a p-group. Similarly, there are slightly different arguments for these two cases in the proof of Theorem 9.1. This dichotomy reflects a division of $\pi(G)$ in **FT**. In **FT**, Feit and Thompson defined π_3 to be the set of all primes p for which $r_p(G) \geq 3$ and some Sylow p-subgroup of G normalizes some nonidentity p'-subgroup of G. They defined π_4 to consist of all other primes p for which $r_p(G) \geq 3$. Our arguments for cases (a) and (b) reflect the arguments in **FT** involving primes in π_3 and π_4.

9. The Uniqueness Theorem

In this section we complete the Uniqueness Theorem (Theorem 9.6). Note that whenever $Y \subseteq X \subset G$ and $X \notin \mathcal{U}$ we have $Y \notin \mathcal{U}$.

Theorem 9.1. Suppose that p is a prime, $M \in \mathcal{M}$, $B \in \mathcal{E}_p(M)$, and B is not cyclic. Assume that

(a) $C_G(b) \subseteq M$ for all $b \in B^{\#}$ or
(b) $\langle \mathcal{U}_G(B; p') \rangle \subseteq M$.

Then $B \in \mathcal{U}$.

Proof. If (a) holds, then for each $K \in \mathcal{U}_G(B; p')$

$$K = \langle C_K(b) \mid b \in B^{\#} \rangle \subseteq M$$

by Proposition 1.16. Therefore (b) also holds and it suffices to prove (a).

We will assume that $B \notin \mathcal{U}$ and obtain a contradiction. Take any $H \in \mathcal{M}(B)$ such that $|H \cap M|_p$ is maximal subject to $H \neq M$. Let R be a Sylow p-subgroup of $H \cap M$ that contains B and let P be a Sylow p-subgroup of M that contains R.

Suppose $K \in \mathcal{U}_G(P; p')$. As $B \subseteq P$, we know from (b) that $K \subseteq M$. Since $[K, P] \subseteq K$, it follows that K centralizes $\mathcal{O}_{p',p}(M)/\mathcal{O}_{p'}(M)$. Hence it follows from Proposition 1.15(a), with G and T replaced by $M/\mathcal{O}_{p'}(M)$ and $\mathcal{O}_{p',p}(M)/\mathcal{O}_{p'}(M)$, that $K \subseteq \mathcal{O}_{p'}(M)$. Since $\mathcal{O}'_p(M)$ itself belongs to $\mathcal{U}_G(P; p')$, this shows that

(9.1) $$\langle \mathcal{U}_G(P; p') \rangle = \mathcal{O}_{p'}(M).$$

By Theorem 6.2, $\mathcal{O}_{p'}(M)Z(J(P)) \lhd M$. Therefore, by the Frattini argument,

(9.2) $$M = \mathcal{O}'_p(M)Z(J(P))N_M(Z(J(P))) = \mathcal{O}'_p(M)N_M(Z(J(P))).$$

It follows that if $\mathcal{O}_{p'}(M) = 1$, then $N_G(P) \subseteq N_G(Z(J(P))) = M$. On the other hand, if $\mathcal{O}_{p'}(M) \neq 1$, then, by (9.1), $N_G(P)$ normalizes $\mathcal{O}_{p'}(M)$ and hence $N_G(P) \subseteq N_G(\mathcal{O}_{p'}(M)) = M$. Thus, in both cases,

(9.3) $$N_G(P) \subseteq M.$$

Consequently P is a Sylow p-subgroup of G.

If $R = P$, then R is a Sylow p-subgroup of G and of H, and, by (9.3), $N_G(R) \subseteq M$. If $R \subset P$, then

$$|H \cap M|_p = |R| < |N_P(R)| \leq |N_G(R) \cap M|_p,$$

and hence, by the choice of H, we know that $\{M\} = \mathcal{M}(N_G(R))$ and $|N_H(R)|_p = |H \cap N_G(R)|_p \leq |H \cap M|_p = |R|$. Consequently, in both cases, R is a Sylow p-subgroup of H and

(9.4) $$N_G(R) \subseteq M.$$

Since, by (b), $\mathcal{O}_{p'}(H) \subseteq M$, we have

(9.5) $$F(H) \subseteq \mathcal{O}_p(H) \times \mathcal{O}_{p'}(F(H)) \subseteq R\mathcal{O}_{p'}(H) \subseteq M.$$

Hence $F(H) \notin \mathcal{U}$ and no subgroup of $F(H)$ lies in $\mathcal{U}$. By Theorem 8.1, $r(F(H)) \leq 2$. Therefore, by Theorem 4.20, $H' \subseteq F(H)$. By (9.5),

$$H' \subseteq F(H) \subseteq R\mathcal{O}_{p'}(H) \subseteq H.$$

Consequently $R\mathcal{O}_{p'}(H) \lhd H$, and, by the Frattini argument and (9.4),

$$H = R\mathcal{O}_{p'}(H)N_H(R) \subseteq \langle \mathcal{O}_{p'}(H), N_G(R) \rangle \subseteq M,$$

which is false. This contradiction completes the proof of the theorem. $\square$

Corollary 9.2. Suppose that $L \in \mathscr{U}$, K is a subgroup of $C_G(L)$, and $r(K) \geq 2$. Then $K \in \mathscr{U}$.

Proof. Let $\mathscr{M}(L) = \{H\}$. Take $B \in \mathcal{E}_p{}^2(K)$ for some prime p. For each $b \in B^{\#}$ we have $C_G(b) \supseteq L$ and hence $C_G(b) \subseteq H$. By Theorem 9.1, $\mathscr{M}(B) = \{H\}$. Therefore $\mathscr{M}(K) = \{H\}$. $\square$

Corollary 9.3. Suppose p is a prime, A is an abelian p-subgroup of G, and B is a noncyclic p-subgroup of G. Assume that $A \in \mathscr{U}$, $\mathrm{m}(A) \geq 3$, and $r_p(C_G(B)) \geq 3$. Then $B \in \mathscr{U}$.

Proof. Let P be a Sylow p-subgroup of G that contains A and take any $B^* \in \mathcal{E}_p{}^3(C_G(B))$. Replacing A and B^* by conjugates, if necessary, we can assume that $B^* \subseteq P$.

By Lemma 4.5, P contains a noncyclic normal subgroup D of order p^2. Therefore $A/C_A(D)$ and $B^*/C_{B^*}(D)$ are cyclic. Hence

$$\mathrm{m}(C_A(D)) \geq 2 \text{ and } \mathrm{m}(C_{B^*}(D)) \geq 2.$$

By hypothesis, $A \in \mathscr{U}$. By successive applications of Corollary 9.2, $\mathscr{U}$ contains $C_A(D)$, D, $C_{B^*}(D)$, B^* and B. $\square$

Lemma 9.4. Suppose that p is a prime, $M \in \mathscr{M}$, and $r_p(F(M)) \geq 3$. Then $\mathscr{U}$ contains every abelian p-group of rank at least three.

Proof. If $F(M)$ is a p-group, then the conclusion follows from Theorem 8.1 and Corollary 9.3. Assume that $F(M)$ is not a p-group. Take any subgroup $A_0 \in \mathcal{E}_p{}^*(F(M))$ such that $\mathrm{m}(A_0) \geq 3$. By Theorem 8.1 and Corollary 9.2, $A_0 \in \mathscr{U}$. The conclusion now follows from Corollary 9.3. $\square$

Lemma 9.5. Suppose p is a prime and $A \in \mathrm{SCN}_3(p)$. Then $A \in \mathscr{U}$.

Proof. Assume that $A \notin \mathscr{U}$. We will obtain a contradiction.
   Take $M \in \mathscr{M}(C_G(A))$ and let $F = F(M)$. By Lemma 9.4,

$$(9.6) \qquad\qquad r_p(F) \leq 2.$$

Choose a prime q as follows: if $r(F) \leq 2$, let q be the largest prime divisor of $|M|$; if $r(F) \geq 3$, let q be some prime for which $r_q(F) \geq 3$. By Theorem 4.20(c), $\mathcal{O}_q(M)$ is a Sylow q-subgroup of M if $r(F) \leq 2$. Since $M = N_G(\mathcal{O}_q(M))$,

$$(9.7) \qquad \mathcal{O}_q(M) \text{ is a Sylow } q\text{-subgroup of } G \text{ if } r(F) \leq 2.$$

Hence, by (9.6), $q \neq p$ if $r(F) \leq 2$. Moreover, by (9.6) and our choice of q, we have $q \neq p$ if $r(F) \geq 3$. Thus $q \neq p$ in both cases.

Let P be a Sylow p-subgroup of $N_G(A)$ and let R be a subgroup of $P \cap M$ that contains A. Then R normalizes $\mathcal{O}_q(M)$. Choose $x \in N_G(R)$ and take $Q \in \mathcal{U}_G^*(R; q)$ such that $\mathcal{O}_q(M) \subseteq Q$. If $r(F) \geq 3$, then, by Lemma 9.4,

$$(9.8) \qquad\qquad Q \subseteq N_G(Q) \subseteq M.$$

On the other hand, if $r(F) \leq 2$, then (9.7) implies that $Q = \mathcal{O}_q(M) \lhd M$ and hence (9.8) is valid in both cases.

Since R is a p-group, $A \lhd\lhd R$. By Theorems 7.6 and 7.4, $\mathcal{O}_{p'}(C_G(R))$ acts transitively on $\mathcal{U}_G^*(R; q)$ by conjugation. Since

$$Q^x \in \mathcal{U}_G^*(R; q) \text{ and } C_G(R) \subseteq C_G(A) \subseteq M,$$

$Q^x = Q^y$ for some $y \in M$. By (9.8),

$$xy^{-1} \in N_G(Q) \subseteq M \text{ and } x = (xy^{-1})y \in M.$$

Thus $N_G(R) \subseteq M$. By taking $R = A$, we have $P \subseteq N_G(A) \subseteq M$. By taking $R = P$, we have

$$(9.9) \qquad\qquad N_G(P) \subseteq M.$$

Let $P_0 = [P, N_G(P)]$ and $D = \mathcal{O}_{p'}(F)$. Since G does not have a normal p-complement, Theorem 1.18 implies that $P_0 \neq 1$.

Suppose that P_0 does not centralize D. By Proposition 1.16,

$$D = \langle\, C_D(B) \mid B \subseteq \Omega_1(A) \text{ and } \Omega_1(A)/B \text{ is cyclic.} \,\rangle$$

Take $B \subseteq \Omega_1(A)$ such that $\Omega_1(A)/B$ is cyclic and P_0 does not centralize $C_D(B)$. Since $A \in \mathrm{SCN}_3(p)$, B is not cyclic. Since $A \notin \mathcal{U}$, it follows that $B \notin \mathcal{U}$. By Theorem 9.1, there exist $y \in B^{\#}$ and $L \in \mathcal{M}$ such that

$$C_G(y) \subseteq L \text{ and } L \neq M.$$

Note that we chose M to be an arbitrary element of $\mathcal{M}(C_G(A))$. Since $C_G(A) \subseteq C_G(y) \subseteq L$, we can apply (9.9), with L in place of M, to conclude that $N_G(P) \subseteq L$. Hence

$$(9.10) \qquad N_G(P) \subseteq L \cap M \text{ and } P_0 \subseteq (N_G(P))' \subseteq (L \cap M)'.$$

Since $D \cap L \subseteq M \cap L$, no subgroup of $D \cap L$ lies in $\mathcal{U}$. As $D = \mathcal{O}_{p'}(F(M))$, Lemma 9.4 implies that $r(D \cap L) \leq 2$. Thus, by (9.10) and Corollary 4.19, P_0 centralizes every chief factor U/V of $L \cap M$ for which $U \subseteq D \cap L$. Therefore, as $D \cap L$ is a p'-group, Lemma 1.9 shows that P_0 centralizes $D \cap L$. However

$$D \cap L \supseteq D \cap C_G(y) \supseteq C_D(B),$$

which is not centralized by P_0. This contradiction shows that P_0 centralizes D, that is,

$$(9.11) \qquad\qquad P_0 \text{ centralizes } \mathcal{O}_{p'}(F).$$

We claim that

(9.12) $$\{M\} = \mathcal{M}(N_G(P_0)).$$

If $r(F) \geq 3$, this follows from (9.11), (9.6), and Lemma 9.4. Suppose that $r(F) \leq 2$. By Theorem 4.20, $M' \subseteq F$. By (9.9), $P \subseteq M$. Since M/M' is abelian,

$$FP \lhd M \text{ and } M = FPN_M(P) = \mathcal{O}_{p'}(F)PN_M(P) = \mathcal{O}_{p'}(F)N_M(P).$$

Since $P_0 = [P, N_G(P)] \lhd N_G(P)$, (9.11) implies that $P_0 \lhd M$, which yields (9.12). Thus (9.12) holds in all cases.

Since $A \notin \mathcal{U}$, it follows that $\Omega_1(A) \notin \mathcal{U}$. By Theorem 9.1, there exists $x \in \Omega_1(A)^{\#}$ such that $C_G(x) \not\subseteq M$. Take $M^* \in \mathcal{M}(C_G(x))$. Since M was chosen arbitrarily from $\mathcal{M}(C_G(A))$, (9.12) yields

$$\{M^*\} = \mathcal{M}(N_G(P_0)) = \{M\},$$

a contradiction. This completes the proof of Lemma 9.5. $\quad\square$

Theorem 9.6 (The Uniqueness Theorem). Suppose that $K \subseteq G$ and $r(K) \geq 2$. Assume that $r(K) \geq 3$ or $r(C_G(K)) \geq 3$. Then $K \in \mathcal{U}$. In particular, if $A \in \mathcal{E}^2(G) - \mathcal{E}^*(G)$, then $A \in \mathcal{U}$.

Proof. By Corollary 9.2, it suffices to prove the result when $r(K) \geq 3$. Then $r_p(K) \geq 3$ for some prime p. Take $B \in \mathcal{E}_p^{\,3}(K)$ of order p^3. Let P be a Sylow p-subgroup of G that contains B. By Lemma 5.1, there exists $A \in \mathrm{SCN}_3(P)$. By Lemma 9.5, $A \in \mathcal{U}$. Since $B \subseteq C_G(B)$, Corollary 9.3 implies that $B \in \mathcal{U}$. Therefore $K \in \mathcal{U}$, as desired. $\quad\square$

CHAPTER III

Maximal Subgroups

$\mathbf{A}$s mentioned in the preface, the proof of the Feit-Thompson Theorem is similar in broad outline to the proof for the special case of CN-groups. There, the CN-group hypothesis yields immediately that the maximal subgroups are Frobenius groups or "three step groups" under a definition different from our definition of three step groups (given in a remark before Proposition 16.1). In contrast, here we have no preliminary restrictions on a maximal subgroup M other than its being solvable. However, having proved the Uniqueness Theorem, we are able to show in fairly short order that M has p-length one for every prime p (Theorem 10.6). Eventually we show that either M is "of Frobenius type" ("almost" a Frobenius group), or M is a three step group (as defined in these notes) (Theorem I). (Incidently, we can obtain Burnside's $p^a q^b$ theorem for odd primes p and q very easily now, as shown in the remark after the proof of Theorem 10.2).

In this chapter we attain part of our final goal by focusing our attention on a single maximal subgroup M. We introduce two normal Hall subgroups M_α and M_σ of M and study their properties in Section 10. The subgroup M_σ plays a role analogous to that of the Fitting subgroup (i.e., the Frobenius kernel) in a Frobenius group. Indeed, if M is a Frobenius group, then $M_\sigma = F(M)$ and $r(M/M_\sigma) = 1$. In Section 11 we study the structure of M under a particular restriction. In Sections 12 and 13 we again allow M to be arbitrary and we study the structure of a complement E to M_σ in M and the embedding of E in M and G.

10. The Subgroups M_α and M_σ

Although the Uniqueness Theorem gives information only about intersections of distinct maximal subgroups, it has powerful consequences for the internal structure of a single maximal subgroup M, as we will see in this section. Among other things, we will show that M contains a normal subgroup M_σ that is a Hall subgroup of G (and thus of M) and has quotient

M/M_σ of rank at most 2 (Theorem 10.2). We will also prove that M has p-length one for all primes p (Theorem 10.6) and determine the possible structures of Sylow p-subgroups of rank at most 2 (Corollary 10.7(b)).

We begin with some notation to be used for the remainder of the proof.

We say that a prime p is *ideal* if $r_p(G) \geq 3$ and $\mathcal{E}^2(P) \cap \mathcal{E}^*(P)$ is empty for every Sylow p-subgroup of G. Equivalently, by Theorem 5.3, for p an ideal prime, the Sylow p-subgroups of G are not narrow. Note that this forces $\mathcal{E}_p{}^2(G) \cap \mathcal{E}_p{}^*(G)$ to be empty.

For a maximal subgroup M of G let

$$\alpha(M) = \{\, p \in \pi(M) \mid r_p(M) \geq 3 \,\},$$
$$\beta(M) = \{\, p \in \alpha(M) \mid p \text{ is ideal} \,\},$$
$$\sigma(M) = \{\, p \in \pi(M) \mid N_G(P) \subseteq M \text{ for some}$$
$$\text{Sylow } p\text{-subgroup } P \text{ of } M\},$$

$$M_\alpha = \mathcal{O}_{\alpha(M)}(M),$$
$$M_\beta = \mathcal{O}_{\beta(M)}(M),$$
$$M_\sigma = \mathcal{O}_{\sigma(M)}(M),$$
$$F_\sigma(M) = \mathcal{O}_{\sigma(M)}(F(M)), \text{ and}$$
$$F_{\sigma'}(M) = \mathcal{O}_{\sigma(M)'}(F(M)).$$

Clearly, by the Uniqueness Theorem (Theorem 9.6),

$$\beta(M) \subseteq \alpha(M) \subseteq \sigma(M) \text{ and } M_\beta \subseteq M_\alpha \subseteq M_\sigma.$$

In some sense the primes in $\beta(M)$, if any, are the best primes in $\pi(M)$. We will show that M_σ is a Hall $\sigma(M)$-subgroup of M and of G and likewise for M_α and M_β (Theorem 10.2 and Lemma 10.8). Moreover, $M_\sigma \neq 1$ and, unless M is "small" in the sense that $r(M) \leq 2$, we have $M_\alpha \neq 1$ (Theorem 10.2). Eventually we will see that M_σ is close to being nilpotent in the sense that $M_\sigma/F(M_\sigma)$ is abelian (Theorem 15.2(g)). We will make relatively little use of M_β.

Note that for $p \in \sigma(M)$ and P any Sylow p-subgroup of M, we have $N_G(P) \subseteq M$, and hence P is a Sylow p-subgroup of G.

Theorem 10.1. Suppose $M \in \mathcal{M}$, $p \in \sigma(M)$, and X is a nonidentity p-subgroup of G.

(a) If $X \subseteq M$, $g \in G$, and $X^g \subseteq M$, then $g = cm$ for some $c \in C_G(X)$ and $m \in M$.

(b) The subgroup $C_G(X)$ acts transitively by conjugation on the set $\{\, M^g \mid g \in G \text{ and } X \subseteq M^g \,\}$.

(c) If $X \subseteq M$, then $N_G(X) = N_M(X)C_G(X)$.

(d) If X is a Sylow p-subgroup of M, $g \in G$, and $X^g \subseteq M$, then $g \in M$.

(e) If $C_G(X) \subseteq M$, $g \in G$, and $X^g \subseteq M$, then $g \in M$.

Proof. We first consider (d). Here both X and X^g are Sylow p-subgroups of M. Hence $(X^g)^h = X$ for some $h \in M$. Then

$$gh \in N_G(X) \subseteq M \text{ and } g = (gh)h^{-1} \in M,$$

as desired.

Now assume (b) temporarily. Then, under the hypothesis of (a),

$$X \subseteq M^{g^{-1}}, \ M^{g^{-1}} = M^c \text{ for some } c \in C_G(X), \text{ and } cg \in N_G(M) = M.$$

Hence $g = c^{-1}(cg)$, $c^{-1} \in C_G(X)$, and $cg \in M$. This proves (a).

Clearly (a) and (b) yield (c) and (e), respectively, as corollaries.

The discussion above shows that (d) is valid and that the remainder of the theorem will follow from (b). We prove (b) by contradiction.

Assume (b) is false and take X to be a counterexample of maximal order. Let $L = N_G(X)$ and take

$$M_1, M_2 \in \{\, M^g \mid g \in G \text{ and } X \subseteq M^g \,\}$$

such that

(10.1) $$M_1{}^c \neq M_2 \text{ for every } c \in C_G(X).$$

We describe the subgroups to be used in our proof by a diagram:

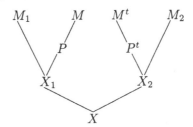

Now $M_2{}^g = M_1$ for some $g \in G$. Thus $X, X^g \subseteq M_1$. If X is a Sylow p-subgroup of M_1, then, by (d), $g \in M_1$, which is impossible. Therefore X is not a Sylow p-subgroup of M_1. It follows that $X \subset X_1$ for some Sylow p-subgroup X_1 of $L \cap M_1$. Similarly $X \subset X_2$ for some Sylow p-subgroup X_2 of $L \cap M_2$.

Let P be a Sylow p-subgroup of L that contains X_1. Take $t \in L$ such that $X_2 \subseteq P^t$. As $p \in \sigma(M)$, M contains a Sylow p-subgroup of G. Replacing M by a conjugate if necessary, we can assume that $P \subseteq M$. Now we have introduced all the subgroups and verified all the containments indicated in the diagram.

Since $X \subset X_1 \subseteq M_1 \cap M$, our maximal choice of X implies that M_1 and M are conjugate under $C_G(X_1)$ and hence under $C_G(X)$. Similarly, M^t and M_2 are conjugate under $C_G(X)$. Therefore, by (10.1),

(10.2) $$M \text{ and } M^t \text{ are not conjugate under } C_G(X).$$

Suppose $r(P) \geq 3$. Then by the Uniqueness Theorem (Theorem 9.6), $P \in \mathscr{U}$. Since $P \subseteq L \cap M$, it follows that $L \subseteq M$. As $t \in L$, this contradicts (10.2). Thus $r(P) \leq 2$. Since P is a Sylow p-subgroup of L, we can conclude from Theorem 4.18(e) that P is a Sylow p-subgroup of $\mathcal{O}_{p',p}(L)$. Thus

$$L = N_L(P)\mathcal{O}_{p',p}(L) = N_L(P)P\mathcal{O}_{p'}(L) = N_L(P)\mathcal{O}_{p'}(L).$$

Hence $t = uv$ for some $u \in N_L(P)$ and $v \in \mathcal{O}_{p'}(L)$.

Since $L = N_G(X)$ and $\mathcal{O}_{p'}(L) \cap X = 1$, we have $v \in \mathcal{O}_{p'}(L) \subseteq C_G(X)$. Since $X \subset P$, the maximal choice of X yields (b) and hence (c) with P in place of X. Thus

$$N_G(P) = N_M(P)C_G(P).$$

Therefore $u = wx$ for some $w \in N_M(P)$ and $x \in C_G(P) \subseteq C_G(X)$. Now

$$w \in M, \quad xv \in C_G(X), \quad t = uv = wxv, \quad \text{and} \quad M^t = M^{wxv} = M^{xv},$$

contrary to (10.2). This completes the proof of Theorem 10.1. $\square$

Theorem 10.2. Let $M \in \mathscr{M}$. Then

(a) M_α is a Hall $\alpha(M)$-subgroup of M and of G,
(b) M_σ is a Hall $\sigma(M)$-subgroup of M and of G,
(c) $M_\alpha \subseteq M_\sigma \subseteq M'$,
(d) $r(M/M_\alpha) \leq 2$ and M'/M_α is nilpotent, and
(e) $M_\sigma \neq 1$.

Proof. Let $M(\alpha)$ be a Hall $\alpha(M)$-subgroup of M. For every nontrivial Sylow subgroup P of $M(\alpha)$, we have $r(P) \geq 3$. Hence, by the Uniqueness Theorem (Theorem 9.6), P is a Sylow subgroup of G and $N(P) \subseteq M$. Thus $\alpha(M) \subseteq \sigma(M)$ and $M(\alpha)$ is contained in some Hall $\sigma(M)$-subgroup $M(\sigma)$ of G.

Now take any $p \in \sigma(M)$ and let P be a Sylow p-subgroup of M. Then $N_G(P) \subseteq M$. Therefore P is a Sylow p-subgroup of G. Clearly $P \cap G' = P$. By the Focal Subgroup Theorem (Theorem 1.17),

$$P = \langle x^{-1}y \mid x, y \in P \text{ and } x \text{ is conjugate to } y \text{ in } G \rangle, \text{ and}$$

$$P \cap M' = \langle x^{-1}y \mid x, y \in P \text{ and } x \text{ is conjugate to } y \text{ in } M \rangle.$$

However, whenever $x \in P$, $g \in G$, and $x^g \in P$, we have $x^g = x^m$ for some $m \in M$ by Theorem 10.1 (with $X = \langle x \rangle$). Hence $P = P \cap M' \subseteq M'$. As p was chosen arbitrarily in $\sigma(M)$,

(10.3) $M(\alpha) \subseteq M(\sigma) \subseteq M'.$

Since $M_\alpha = \mathcal{O}_{\alpha(M)}(M)$, the section $F(M/M_\alpha)$ is an $\alpha(M)'$-group and is isomorphic to a subgroup of M. Hence, by the definition of $\alpha(M)$, we have $r(F(M/M_\alpha)) \leq 2$. Clearly $M_\alpha \subseteq M(\alpha)$. By (10.3) and Theorem 4.20,

$$M(\sigma)/M_\alpha \subseteq M'/M_\alpha = (M/M_\alpha)' \subseteq F(M/M_\alpha).$$

Thus $M(\sigma)/M_\alpha$ is an $\alpha(M)'$-group. By (10.3), this proves that $M(\alpha) = M_\alpha$, which yields (a). Since $M(\sigma)$ is a Hall subgroup of M, it follows that $M(\sigma)/M_\alpha$ is a Hall subgroup of $F(M/M_\alpha)$ and is normal in M/M_α. Consequently $M(\sigma) \lhd M$, which shows that $M(\sigma) = M_\sigma$. Thus (b) holds, and (10.3) yields (c).

By the definition of $M(\alpha)$,

$$\mathrm{r}(M/M_\alpha) = \mathrm{r}(M/M(\alpha)) \le 2.$$

As mentioned above, $M'/M_\alpha \subseteq F(M/M_\alpha)$, so we obtain (d).

Now we have proved all parts of the conclusion except (e). To obtain (e), we may assume that $M_\alpha = 1$. Then $\mathrm{r}(M) \le 2$. Let q be the largest prime divisor of $|M|$. By Theorem 4.20, $\mathcal{O}_q(M)$ is a Sylow q-subgroup of M. Clearly $N_G(\mathcal{O}_q(M)) = M$. Hence $q \in \sigma(M)$ and $M_\sigma \ne 1$. This completes the proof of the theorem. $\square$

Remark. A famous theorem of Burnside asserts that all groups of order $p^a q^b$ (for some primes p and q) are solvable. Burnside's original proof relies heavily on character theory. In 1970, David Goldschmidt published a short character-free proof for the case when p and q are both odd [13]. During the 1975 class mentioned in the preface, David T. Price pointed out that at this stage of our proof we can easily verify Burnside's result for p and q odd by using some of Goldschmidt's methods. We do so now.

Assume $\pi(G) = \langle p, q \rangle$ for some primes p and q. Since p-groups are nilpotent, $p \ne q$. As $|G|_p \ne |G|_q$, we can assume that $|G|_p > |G|_q$. Let P and P_1 be two distinct Sylow p-subgroups of G, chosen such that $|P \cap P_1|$ is as large as possible. Let $R = P \cap P_1$. Since

$$|P : R| \le |G : P_1| = |G|_q < |P|,$$

we know $R \ne 1$. Take $M \in \mathcal{M}(N_G(R))$. By Theorem 10.2, M_σ is a nonidentity Hall subgroup of M and of G and is normal in M. Hence M_σ is a Sylow p-subgroup or a Sylow q-subgroup of G.

Suppose M_σ is a Sylow p-subgroup of G. Since $R \subset N_P(R) \subseteq M_\sigma \cap P$, we know, by our choice of P and P_1, that $M_\sigma = P$. But then, similarly, $M_\sigma = P_1$, a contradiction. Hence M_σ must be a Sylow q-subgroup of G.

Now $G = M_\sigma P = MP$. Let

$$N = \bigcap_{x \in G} M^x.$$

Then $N \lhd G$ and $N \subseteq M$. Therefore $N = 1$. However,

$$N = \bigcap_{y \in M, z \in P} M^{yz} = \bigcap_{z \in P} M^z \supseteq \bigcap_{z \in P} N_P(R)^z \supseteq Z(P) \supset 1,$$

a contradiction. This completes the proof.

The interested reader can also find a complete character-free proof of Burnside's Theorem, based on ideas of Bender, Goldschmidt, and Matsuyama, in [21, pp. 121-130] and in [18, pp. 11-19].

Lemma 10.3. Suppose $M \in \mathcal{M}$, X is an $\alpha(M)'$-subgroup of M, and $r(C_{M_\alpha}(X)) \geq 2$. Then $C_M(X) \in \mathcal{U}$.

Remark. The Uniqueness Theorem (Theorem 9.6) immediately shows that $C_M(X) \in \mathcal{U}$ if $r(C_{M_\alpha}(X)) \geq 3$. The point of this lemma is to do a little better.

Proof. Take a prime p for which $r_p(C_{M_\alpha}(X)) \geq 2$ and choose a group $B \in \mathcal{E}_p(C_{M_\alpha}(X))$ of maximal order. Now X normalizes M_α and has order relatively prime to $|M_\alpha|$. By Proposition 1.5, X normalizes some Sylow p-subgroup P of M_α that contains B. Clearly we can assume that $B \notin \mathcal{U}$.

By the Uniqueness Theorem, $r(C_P(B)) \leq 2$. Hence

$$|B| = p^2 \text{ and } \Omega_1(C_P(B)) = B \subseteq C_P(X).$$

By Corollary 1.12, $C_P(X) = P$. Therefore $r(C_M(X)) \geq r(P) \geq 3$ because $p \in \alpha(M)$. By the Uniqueness Theorem, $C_M(X) \in \mathcal{U}$. $\square$

Lemma 10.4. Suppose $M \in \mathcal{M}$, $p \in \pi(M)$, and P is a Sylow p-subgroup of M.

(a) If p divides $|M/M'|$, then $p \notin \sigma(M)$.

(b) Assume $p \notin \sigma(M)$ and $M_\alpha \neq 1$. Then there exists $x \in \Omega_1(Z(P))^\#$ such that $\{M\} \neq \mathcal{M}(C_G(x))$ and $C_{M_\alpha}(x)$ is a Z-group.

(c) Assume $p \notin \sigma(M)$ and $r_p(M) = 2$. Then p is not ideal and $\mathcal{E}_p^2(M) \subseteq \mathcal{E}_p^*(G)$.

Proof. (a) This follows immediately from Theorem 10.2(c).

(b) Assume that we have a counterexample. Choose $u \in N_G(P) - M$ and $y \in \Omega_1(Z(P))^\#$. Then either $C_{M_\alpha}(y)$ is not a Z-group or $\{M\} = \mathcal{M}(C_G(y))$. By Lemma 10.3, $\{M\} = \mathcal{M}(C_G(y))$ in either case. Similarly, $\{M\} = \mathcal{M}(C_G(y^{u^{-1}}))$ because $y^{u^{-1}} \in Z^\#$. Since $(C_G(y^{u^{-1}}))^u = C_G(y)$, we have $M^u = M$. Thus $u \in N_G(M) = M$, contrary to our choice of u.

(c) Suppose $A \in \mathcal{E}_p^2(M)$ and $A \notin \mathcal{E}_p^*(G)$. Then, by the Uniqueness Theorem, $A \in \mathcal{U}$. Let P be a Sylow p-subgroup of G that contains A. Then $N_G(P) \subseteq M$ because $A \subseteq N_G(P)$. However, we assumed that $p \notin \sigma(M)$. This contradiction shows that $\mathcal{E}_p^2(M) \subseteq \mathcal{E}_p^*(G)$.

Since $r_p(M) = 2$, $\mathcal{E}_p^2(M)$ is not empty. Hence $\mathcal{E}_p^2(M) \cap \mathcal{E}_p^*(G)$ is also not empty. By the definition of an ideal prime, p is not ideal. This completes the proof of the lemma. $\square$

Lemma 10.5. Suppose that $M \in \mathcal{M}$, $p \in \sigma(M)'$, $X \in \mathcal{E}_p^1(G)$, and $N_G(X) \subseteq M$. Then $r_p(M) = 2$, p is not ideal, and there exists $A \in \mathcal{E}_p^2(G)$ containing X.

Proof. Since $\alpha(M) \subseteq \sigma(M)$, it follows that $p \notin \alpha(M)$, i.e., $r_p(M) \leq 2$. Let P be a Sylow p-subgroup of M that contains X.

Suppose $r_p(M) = 1$. Then, by Lemma 4.5, P is cyclic. Therefore

$$X = \Omega_1(P) \text{ and } N_G(P) \subseteq N_G(X) \subseteq M,$$

contrary to the assumption that $p \notin \sigma(M)$. Thus $r(P) = r_p(M) = 2$. Similarly $X \neq \Omega_1(Z(P))$, and hence $X\Omega_1(Z(P)) \in \mathcal{E}^2(P)$. By Lemma 10.4, p is not ideal. $\square$

So far, we have obtained only mild restrictions on the structure of the proper subgroups of G. It is conceivable that they may be extremely complicated, subject only to being solvable. The following important result shows that this is not the case.

Recall that a group H has *p-length one* for a given prime p if $H/\mathcal{O}_{p',p}(H)$ is a p'-group.

Theorem 10.6. Suppose p is a prime and H is a proper subgroup of G. Then H has p-length one.

Proof. Take $M \in \mathcal{M}(H)$. It will suffice to prove that M has p-length one. If $r_p(M) \leq 2$, this is a consequence of Theorem 4.18, so assume that $r_p(M) \geq 3$.

Let K be a complement to M_α in M. By Theorem 10.2, $M_\alpha \subseteq M'$. By Lemma 6.3(a), $M_\alpha = [M_\alpha, K]$. Take $q \in \pi(K/K')$ and let Q be a Sylow q-subgroup of K. By Lemma 10.4, there exists $x \in Q$ such that x has order q and $C_{M_\alpha}(x)$ is a Z-group. Therefore, by Theorem 3.6, $[M_\alpha, K]$ has p-length one. Since $M_\alpha = [M_\alpha, K]$, the result follows. $\square$

The theorem above will be convenient for many applications, although its use could be avoided in some of them.

Corollary 10.7. Suppose that p is a prime and P is a Sylow p-subgroup of G.

(a) Take V to be any complement of P in $N_G(P)$. Then we have $P = [P, V] \subseteq N_G(P)'$.

(b) Suppose $r(P) \leq 2$. Then either P is abelian or P is the central product of a nonabelian subgroup P_1 of order p^3 and exponent p and a cyclic subgroup P_2 for which $\Omega_1(P_2) = Z(P_1)$.

(c) Suppose $Q \subseteq P$, $x \in G$, and $Q^x \subseteq P$. Then $Q^x = Q^y$ for some $y \in N_G(P)$.

(d) For every subgroup Q of P, the group $N_P(Q)$ is a Sylow p-subgroup of $N_G(Q)$.

(e) Suppose R is a p-subgroup of G and $Q \subseteq P \cap R$ and $Q \lhd N_G(P)$. Then $Q \lhd N_G(R)$.

Proof. Take $M \in \mathscr{M}(N_G(P))$. Then $p \in \sigma(M)$. By the previous result, $P \subseteq \mathcal{O}_{p',p}(M)$. We will apply Lemma 6.6.

(a) By Theorem 10.2, $P \subseteq M_\sigma \subseteq M'$. Therefore, by Lemma 6.6, we know that $P \subseteq (N_M(P))' = (N_G(P))'$. By Lemma 6.3, $[P, V] = P$.

(b) This follows from (a) and Theorem 4.16.

(c) By Theorem 10.1, there exists $u \in M$ such that $Q^u = Q^x$. By Lemma 6.6(c), there exists $y \in N_M(P)$ such that $Q^y = Q^u$.

(d) Let S be a Sylow p-subgroup of $N_G(Q)$. Take $x \in G$ such that $S^x \subseteq P$. Then $Q^x \subseteq P$. Take y as in (c). Then $Q^{xy^{-1}} = Q$, $S^{xy^{-1}} \subseteq P$, and $S^{xy^{-1}}$ is a Sylow p-subgroup of $N_G(Q)$.

(e) Take $x \in G$ such that $R^x \subseteq P$. Take any $y \in N_G(R)$. Then P contains Q^x and Q^{yx}. By (c) and the assumption that $Q \vartriangleleft N_G(P)$, we have $Q^x = Q = Q^{yx}$. Hence $Q = Q^y$. $\square$

Recall that in the beginning of this section we defined $\beta(M)$ to be the set of all ideal primes in $\alpha(M)$ and M_β to be $\mathcal{O}_{\beta(M)}(M)$. We also noted that a prime p is ideal if and only if the Sylow p-subgroups of G are not narrow.

Lemma 10.8. Let $M \in \mathscr{M}$. Then

    (a) M_β is a Hall $\beta(M)$-subgroup of M and of G,
    (b) M' and M_σ have nilpotent Hall $\beta(M)'$-subgroups, and
    (c) for each prime $p \in \pi(M) - \beta(M)$, both M' and M_σ have normal p-complements and p is the largest prime divisor of $|M/\mathcal{O}_{p'}(M)|$.

Proof. Let $M(\beta)$ be a Hall $\beta(M)$-subgroup of M.

Suppose $p \in \pi(M) - \beta(M)$. Let P be a Sylow p-subgroup of M. Then P is narrow. By Theorem 10.6, M has p-length one. Consequently, by Theorem 5.6, M' has a normal p-complement, which must contain $M(\beta)$, and p is the largest prime divisor of $|M/\mathcal{O}_{p'}(M)|$.

The intersection of the normal p-complements above, as p varies over $\pi(M) - \beta(M)$, is a normal $\beta(M)$-subgroup of M'. It contains $M(\beta)$ and hence must be equal to $M(\beta)$. Therefore

$$M(\beta) = \mathcal{O}_{\beta(M)}(M') \vartriangleleft M \quad \text{and} \quad M(\beta) = \mathcal{O}_{\beta(M)}(M) = M_\beta.$$

This proves (a) and (c). Since M'/M_β has a normal p-complement for every prime divisor p of $|M'/M_\beta|$, M'/M_β is nilpotent. As $M_\beta \subseteq M_\sigma \subseteq M'$, we obtain (b). $\square$

Corollary 10.9. Let $M \in \mathscr{M}$.

    (a) Suppose p and q are distinct primes in $\beta(M)'$ and X is a q-subgroup of M. Assume $X \subseteq M'$ or $p < q$. Then
        (1) X centralizes a Sylow p-subgroup of M_σ,
        (2) if $p \in \alpha(M)$, then $C_M(X) \in \mathscr{U}$,

(3) if X is a Sylow q-subgroup of M', then $N_M(X)'$ contains a Sylow p-subgroup of M'.

(b) If $H \in \mathcal{M} - \{M\}$ and $N_G(S) \subseteq H \cap M$ for some Sylow subgroup S of G, then $M = (H \cap M)M_\beta$ and $\alpha(M) = \beta(M)$.

Proof. (a) Let W be a Hall $\{p, q\}$-subgroup of XM' that contains X. Then W is contained in a Hall $\beta(M)'$-subgroup W^* of XM'. Consequently, by Lemma 10.8(b), $W^* \cap M'$ is nilpotent, so $W \cap M'$ is nilpotent.

We claim that W is nilpotent. This is clear if $X \subseteq M'$, for then $W = W \cap M'$. Now assume $X \not\subseteq M'$. Then, by hypothesis, $p < q$. Therefore, by Lemma 10.8(c), $\mathcal{O}_{p'}(M)$ contains all the q-elements of M. In particular, $W \cap \mathcal{O}_{p'}(M)$ is a normal Sylow q-subgroup of W. Since $W \cap M'$ is nilpotent and $W/(W \cap M') \cong WM'/M' \subseteq XM'/M'$, which is a q-group, $\mathcal{O}_p(W \cap M')$ is a normal Sylow p-subgroup of W. Consequently W is nilpotent.

Since $M_\sigma \lhd M'$, the intersection $W \cap M_\sigma$ is a Hall $\{p, q\}$-subgroup of M_σ. Thus we obtain (1). Clearly (2) follows from (1) by the Uniqueness Theorem.

To obtain (3), assume that X is a Sylow q-subgroup of M'. Then $X = \mathcal{O}_q(W^*)$. Since $M' = M_\beta W^*$,

$$M_\beta X = \mathcal{O}_{\beta(M) \cup \{q\}}(M') \lhd M.$$

Let $U = N_M(X)$. By the Frattini argument, $M = (M_\beta X)U = M_\beta U$. Now $\mathcal{O}_p(W)$ is a Sylow subgroup of M' contained in U and $(|\mathcal{O}_p(W)|, |M_\beta|) = 1$. Therefore, by Lemma 6.5, $\mathcal{O}_p(W) \subseteq U'$. This completes the proof of (3) and of (a).

(b) In this situation, $S \neq 1$. Let q be the unique prime divisor of $|S|$. Then $q \in \sigma(M) - \beta(M)$ and $S \subseteq M_\sigma \subseteq M'$. Hence, by Lemma 10.8(b), $M_\beta S \lhd M$. The proof of (a) shows that $M = M_\beta N_M(S)$. Therefore $M = M_\beta(H \cap M)$. Clearly $N_M(S) \notin \mathcal{U}$. By (a)(2), $\alpha(M) = \beta(M)$. $\square$

Proposition 10.10. Suppose that p and q are distinct prime numbers, $A \in \mathcal{E}_p^2(G) \cap \mathcal{E}_p^*(G)$, and $Q \in \mathcal{M}_G^*(A; q)$. Assume that $q \in \pi(C_G(A))$. Then for some Sylow p-subgroup P of G that contains A,

(a) $N_G(P) = \mathcal{O}_{p'}(C_G(P))(N_G(P) \cap N_G(Q))$,

(b) $P \subseteq N_G(Q)'$, and

(c) if Q is cyclic or $\mathcal{E}^2(Q) \cap \mathcal{E}^*(Q)$ is not empty, then P centralizes Q.

Proof. By Proposition 7.5, A satisfies Hypothesis 7.1. Therefore Theorem 7.3 and Theorem 7.4 yield a Sylow p-subgroup P of G that contains A and satisfies (a).

By Corollary 10.7, $P \subseteq N_G(P)'$. Thus (b) follows from (a) by Lemma 6.5.

Under the hypothesis of (c), $N_G(Q)'/C_{N_G(Q)'}(Q)$ is a q-group by Theorem 5.5(a). Thus P lies in $C_G(Q)$. $\square$

Proposition 10.11. Suppose $M \in \mathscr{M}$ and K is a $\sigma(M)'$-subgroup of M. Then

    (a) $K \notin \mathscr{U}$,
    (b) $\mathrm{r}(C_K(M_\sigma)) \leq 1$,
    (c) $C_K(M_\sigma) \cap M'$ is a cyclic normal subgroup of M, and
    (d) if $p \in \sigma(M)'$, $P \in \mathcal{E}_p^{\,1}(N_M(K))$, $C_{M_\sigma}(P) = 1$, and K is an abelian p'-group, then $[K, P]$ centralizes M_σ and is a cyclic normal subgroup of M.

Proof. (a) Let E be a Hall $\sigma(M)'$-subgroup of M that contains K. Let p be the largest prime divisor of $|E|$. Since $\alpha(M) \subseteq \sigma(M)$, $\mathrm{r}(E) \leq 2$. By Theorem 4.20, $P = \mathcal{O}_p(E)$ is a Sylow p-subgroup of E and hence of M. Since $p \notin \sigma(M)$, we know $N_G(P) \not\subseteq M$. Therefore $K \notin \mathscr{U}$ because

$$K \subseteq E \subseteq M \cap N_G(P).$$

    (b) Suppose $\mathrm{r}_p(C_K(M_\sigma)) \geq 2$ for some prime p. Take $A \in \mathcal{E}_p^{\,2}(C_K(M_\sigma))$ and $q \in \sigma(M)$. Let Q be a Sylow q-subgroup of M_σ. Then $Q \in \mathit{U}_G^{*}(A; q)$, $q \in \pi(C_G(A))$, and $N_G(Q) \subseteq M$.

By (a), $A \notin \mathscr{U}$. Thus, by the Uniqueness Theorem, $\mathrm{r}(C_G(A)) \leq 2$ and $A \in \mathcal{E}_p^{*}(G)$. As $M_\alpha \subseteq M_\sigma \subseteq C_G(A)$, we have $M_\alpha = 1$. By Theorem 10.2, M'/M_α is nilpotent and hence M' is nilpotent.

By Proposition 10.10, some Sylow p-subgroup P of G lies in $N_G(Q)'$ and therefore in M'. Consequently $P = \mathcal{O}_p(M') \lhd M$, $M = N_G(P)$, and $p \in \sigma(M)$. However $p \in \sigma(K) \subseteq \sigma(M)'$. This contradiction completes the proof of (b).

    (c) We can apply (b) to $Z = \mathcal{O}_{\sigma(M)'}(F(M))$ since $[Z, M_\sigma] \subseteq Z \cap M_\sigma = 1$, and it follows that Z is cyclic. Therefore $M' \subseteq C_M(Z)$ and

$$C_K(M_\sigma) \cap M' \subseteq C_M(M_\sigma Z) \subseteq C_M(F(M)) \subseteq F(M).$$

Thus $C_K(M_\sigma) \cap M'$ is contained in Z and is normal in M because Z is a cyclic normal subgroup of M.

    (d) With $K_0 = [K, P]$, we have $K_0^{\,P} = K_0$ and $K = K_0 \times C_K(P)$ because K is an abelian p'-group. Thus P acts on $K_0 M_\sigma$ and satisfies

$$C_{K_0 M_\sigma}(P) = C_{K_0}(P) C_{M_\sigma}(P) = 1.$$

Therefore, by Theorem 3.7, $K_0 M_\sigma$ is nilpotent. Hence K_0 centralizes M_σ, and (c) shows that K_0 is a cyclic normal subgroup of M. $\square$

Lemma 10.12. Suppose $M, H \in \mathscr{M}$ and H is not conjugate to M in G. Then

    (a) $M_\alpha \cap H_\sigma = 1$ and $\alpha(M)$ is disjoint from $\sigma(H)$, and
    (b) if M_σ is nilpotent, then $M_\sigma \cap H_\sigma = 1$ and $\sigma(M)$ is disjoint from $\sigma(H)$.

Proof. Suppose $p \in \sigma(M) \cap \sigma(H)$. Then some Sylow p-subgroup S of G lies in M and in some conjugate H^g of H.

By the Uniqueness Theorem, $r(S) \leq 2$ because $H^g \neq M$. Furthermore, S is not normal in M because $N_G(S) \subseteq H^g$. This shows that $p \notin \alpha(M)$ and that M_σ is not nilpotent.

Now we are done because $\pi(M_\alpha \cap H_\sigma) \subseteq \alpha(M) \cap \sigma(H) \subseteq \sigma(M) \cap \sigma(H)$ and $\pi(M_\sigma \cap H_\sigma) \subseteq \sigma(M) \cap \sigma(H)$. $\square$

Lemma 10.13. Suppose $p \in \pi(G)$, $A \in \mathcal{E}_p^2(G) \cap \mathcal{E}_p^*(G)$, and P is a nonabelian p-subgroup of G that contains A. Let $Z_0 = \Omega_1(Z(P))$ and $A_0 \in \mathcal{E}^1(A)$ such that $A_0 \neq Z_0$. Then

  (a) $Z_0 \in \mathcal{E}^1(A)$,
  (b) $C_P(A) = A_0 \times Z$ with Z a cyclic subgroup that contains Z_0, and
  (c) $N_P(A)$ acts transitively by conjugation on $\mathcal{E}^1(A) - \{Z_0\}$.

Proof. Let S be a Sylow p-subgroup of G that contains P and let $Z_1 = \Omega_1(Z(S))$. Then

(10.4) $A \in \mathcal{E}^2(S) \cap \mathcal{E}^*(S)$, $A = \Omega_1(C_S(A))$, and $r(C_S(A)) = 2$.

Therefore $Z_0, Z_1 \subseteq A$. Clearly $Z_1 \subseteq Z_0$. Thus $Z_1 \subseteq Z_0 \subseteq A$. In particular, $Z_0 = Z_1$ unless $A = Z_0 \subseteq Z(P)$. As P is not abelian, it follows that

(10.5)   if $C_S(A)$ is abelian, then $Z_1 = Z_0 \in \mathcal{E}^1(A)$.

We claim that

(10.6) $Z_0 = Z_1 \in \mathcal{E}^1(A)$ and $C_S(A) = A_0 \times Y$

for some cyclic subgroup Y of S such that $\Omega_1(Y) = Z_1$.

Assume first that $r(S) = 2$. Then, by Corollary 10.7(b), $Z(S)$ is cyclic and $S = S_1 Z(S)$ for some subgroup S_1 of S having order p^3 and exponent p such that $Z(S_1) = \Omega_1(Z(S)) = Z_1$. It is easy to see that

$$A \subset \Omega_1(S) = S_1.$$

By the structure of S_1, we have $C_S(A) = AZ(S)$, which is abelian. Hence, by (10.5), $A = A_0 \times Z_0$, and we obtain (10.6) for $Y = Z(S)$.

Now assume that $r(S) \geq 3$. By (10.4) and Theorem 5.3, S is narrow and $|Z_1| = p$. Let $T = C_S(\Omega_1(Z_2(S)))$ and take any $A_1 \in \mathcal{E}^1(A) - \{Z_1\}$. Then $A = A_1 \times Z_1$ and $C_S(A_1) = C_S(A)$, which has rank two by (10.4). Therefore, by Theorem 5.3, $C_T(A)$ is cyclic and $C_S(A) = A_1 \times C_T(A)$, which is abelian. By (10.5), $Z_0 = Z_1$. Clearly $Z_1 = \Omega_1(C_T(A))$ because $Z_1 \subseteq C_T(A)$. By hypothesis, $A_0 \neq Z_0$. Hence we can assume that A_1 was chosen to be A_0. Now (10.6) follows with $Y = C_T(A)$.

Since (10.6) holds in both cases, we obtain (a) and (b). Moreover, $C_P(A) \subset P$ because $C_P(A)$ is abelian. By (10.4), $A = \Omega_1(C_P(A))$. Thus

$$C_P(A) \subset N_P(C_P(A)) \subseteq N_P(\Omega_1(C_P(A))) = N_P(A).$$

Take $x \in N_P(A) - C_P(A)$. Then x induces an automorphism of order p on A by conjugation, and this automorphism centralizes Z_0. By the structure of A, this forces $\langle x \rangle$ to permute transitively all of the p-subgroups of order p in A other than Z_0. This proves (c) and completes the proof of the lemma. $\square$

Proposition 10.14. Suppose that $p \in \beta(G)$ and P is a Sylow p-subgroup of G.

   (a) The sets $\mathcal{E}^2(P) \cap \mathcal{E}^*(P)$ and $\mathcal{E}_p{}^2(G) \cap \mathcal{E}^*(G)$ are empty.
   (b) Every p-subgroup R of G such that $\mathrm{r}(R) \geq 2$ lies in $\mathscr{U}$.
   (c) If X is a subgroup of P, then $N_P(X) \in \mathscr{U}$.
   (d) For every nonidentity $\beta(M)$-subgroup Y of M, $N_G(Y) \subseteq M$.

Proof. (a) By the definition of $\beta(G)$, we know that $\mathcal{E}^2(P) \cap \mathcal{E}^*(P)$ is empty. We noted earlier that this easily forces $\mathcal{E}_p{}^2(G) \cap \mathcal{E}^*(G)$ to be empty.

(b) We can assume that $R \subseteq P$. Take $A \in \mathcal{E}^2(R)$. By (a), $A \subset B$ for some elementary abelian subgroup B of P. Then $B \subseteq C_G(A)$ and $\mathrm{m}(B) \geq 3$. By the Uniqueness Theorem, $A \in \mathscr{U}$. Hence $R \in \mathscr{U}$.

(c) Let $Q = N_P(X)$. If $\mathrm{r}(Q) \geq 2$, then $Q \in \mathscr{U}$, by (b). Therefore assume that $\mathrm{r}(Q) = 1$. Then Q is cyclic, $X \operatorname{char} Q$, and $N_P(Q) \subseteq N_P(X) = Q$. Since P is nilpotent, this forces $Q = P$, contrary to the assumption that $\mathrm{r}(P) \geq 3$.

(d) Let $q \in \pi(F(Y))$ and $X = \mathcal{O}_q(Y)$. We can assume that $q = p$ and $X \subseteq P \subseteq M$. Then $N_P(X) \subseteq M$ and, by (c),

$$Y \subseteq N_G(X) \subseteq M.$$

This completes the proof of the proposition. $\square$

11. Exceptional Maximal Subgroups

Throughout this section we assume the following notation and conditions.

Hypothesis 11.1. Suppose $M \in \mathcal{M}$, $p \in \sigma(M)'$, $A_0 \in \mathcal{E}_p{}^1(M)$, and

$$N_G(A_0) \subseteq M.$$

By Lemma 10.5, $\mathrm{r}_p(M) = 2$ and $A_0 \subseteq A$ for some $A \in \mathcal{E}_p{}^2(M)$. Then $A \subseteq P$ for some Sylow p-subgroup P of M. As $p \notin \sigma(M)$,

$$N_G(P) \not\subseteq M,$$

and since $C_G(A) \subseteq C_G(A_0) \subseteq M$,

$$A \in \mathcal{E}_p{}^*(G).$$

We will use this choice of A and P and these facts throughout Section 11.

Since we would expect that $p \in \sigma(M)$ whenever $N_G(A_0) \subseteq M$ with $A_0 \in \mathcal{E}_p{}^1(G)$ and, since maximal subgroups H such that $\mathrm{r}(H/H_\sigma) = 1$ seem more natural and easier to work with, we consider M to be an *exceptional maximal subgroup*.

We will show that M_σ is nilpotent, M has abelian Sylow p-subgroups, and $AM_\sigma \lhd M$ (Theorems 11.3, 11.5, and 11.7). In Section 12, we will see that all maximal subgroups H for which $\mathrm{r}(H/H_\sigma) = 2$ are exceptional (Proposition 12.4). In the proof of this proposition and very often thereafter we will consider maximal subgroups $M^* \in \mathcal{M}(N_G(X))$ for $X \in \mathcal{E}_p^{\,1}(M)$ and, in the exceptional case when $X \not\subseteq M^*_\sigma$, we will have the situation described above (with M^* and X in place of M and A_0). This is the reason for studying exceptional maximal subgroups.

The main results of this section will be generalized in Theorem 12.5 and there will be no reference to Section 11 thereafter.

Finally, we note that the following proofs depend heavily on the ideas underlying the fundamental Thompson Transitivity Theorem (Theorem 7.6).

Lemma 11.1. Suppose that $g \in G - M$, $A \subseteq M^g$, $q \in \sigma(M)$, and that Q_1 and Q_2 are A-invariant Sylow q-subgroups of M_σ and $M_\sigma{}^g$, respectively. Then

    (a) $Q_1 \cap Q_2 = 1$, and
    (b) if $X \in \mathcal{E}^1(A)$, then $C_{Q_1}(X) = 1$ or $C_{Q_2}(X) = 1$.

Proof. By Proposition 7.5, the subgroup A satisfies Hypothesis 7.1 because $A \in \mathcal{E}_p{}^*(G)$. Thus, if (a) or (b) is false, Lemma 7.1 shows that $Q_2 = Q_1{}^k$ for some element $k \in C_G(A)$. Then $Q_2 \subseteq M$ because $C_G(A) \subseteq M$. Therefore, by Theorem 10.1(d), $g \in M$, a contradiction. $\square$

Corollary 11.2. Suppose $g \in G - M$ and $A \subseteq M^g$. Then

    (a) $M_\sigma \cap M^g = 1$, and
    (b) $M_\sigma \cap C_G(A_0{}^g) = 1$.

Proof. Suppose (a) is false. Take $q \in \pi(M_\sigma \cap M^g)$. Then $q \in \sigma(M)$. Since $M_\sigma \cap M^g = M_\sigma \cap M_\sigma{}^g$ and A normalizes M_σ and $M_\sigma{}^g$, Proposition 1.5 shows that A normalizes Sylow q-subgroups Q_0, Q_1, and Q_2 of $M_\sigma \cap M^g$, M_σ, and $M_\sigma{}^g$, respectively, such that $Q_0 \subseteq Q_1 \cap Q_2$.

By Lemma 11.1, $Q_1 \cap Q_2 = 1$, a contradiction. This proves (a). For (b), note that $C_G(A_0{}^q) \subseteq M^g$. $\square$

Theorem 11.3. The group M_σ is nilpotent.

Proof. Take $g \in N_G(P) - N_M(P)$. Then $A_0{}^g \subseteq P \subseteq M$, and $A_0{}^g$ acts in a fixed-point free manner on M_σ by Corollary 11.2(b). Therefore, by Theorem 3.7, M_σ is nilpotent. $\square$

Corollary 11.4. Suppose $H \in \mathcal{M}(A)$ and $M_\sigma \cap H_\sigma \neq 1$. Then $M = H$.

Proof. Since M_σ is nilpotent, Lemma 10.12(b) shows that H is conjugate to M, say $H = M^g$ for $g \in G$. Then, by Corollary 11.2(a), $g \in M$ because $M_\sigma \cap M^g \neq 1$. $\square$

Theorem 11.5. The Sylow p-subgroups of M are abelian.

Proof. Suppose P is not abelian. Take $g \in N_G(P) - N_M(P)$ and $q \in \sigma(M)$, and let Q_1 be a P-invariant Sylow q-subgroup of M_σ. By Proposition 1.16,

$$Q_1 = \langle C_{Q_1}(X) \mid X \in \mathcal{E}^1(A) \rangle$$

and similarly for $Q_2 = Q_1{}^g$.

Thus there exist subgroups X_1, $X_2 \in \mathcal{E}^1(A)$ such that $C_{Q_1}(X_1) \neq 1$ and $C_{Q_2}(X_2) \neq 1$.

By Lemma 11.1(b), each subgroup $X \in \mathcal{E}^1(A)$ satisfies $C_{Q_1}(X) = 1$ or $C_{Q_2}(X) = 1$. Therefore $C_{Q_1}(X_2) = 1$, and hence X_2 is not conjugate to X_1 in P.

By Lemma 10.13(c), all of the subgroups in $\mathcal{E}^1(A) - \{X\}$, where $X = \Omega_1(Z(P))$, are conjugate in P. Consequently $X = X_1$ or $X = X_2$. However, because $X = X^g$ and hence $C_{Q_2}(X) = C_{Q_1}(X)^g$, we have $C_{Q_1}(X) = C_{Q_2}(X) = 1$, a contradiction. $\square$

Corollary 11.6. We have

   (a) $A = \Omega_1(P)$,
   (b) $C_{M_\sigma}(A) = 1$, and
   (c) there exist subgroups A_1, $A_2 \in \mathcal{E}_p{}^1(A)$ such that $A_1 \neq A_2$ and $C_{M_\sigma}(A_1) = C_{M_\sigma}(A_2) = 1$.

Proof. Since $A \in \mathcal{E}^2(P)$ and P is abelian of rank two,

$$A = \Omega_1(P) \triangleleft N_G(P).$$

As $N_G(P) \not\subseteq M$ and G has odd order,

$$|N_G(P) : N_M(P)| \geq 3.$$

Thus we can find elements g_1, $g_2 \in N_G(P) - N_M(P)$ such that $g_1 g_2^{-1} \notin M$. Then $g_1 g_2^{-1} \notin N_G(A_0)$ because $N_G(A_0) \subseteq M$. Therefore $A_0{}^{g_1} \neq A_0{}^{g_2}$. Since $A_0 \subseteq A \triangleleft N_G(P)$, it follows that $A_1 = A_0{}^{g_1}$ and $A_2 = A_0{}^{g_2}$ lie in A. By Corollary 11.2(b), $C_{M_\sigma}(A_1) = C_{M_\sigma}(A_2) = 1$. In particular, $C_{M_\sigma}(A) = 1$. $\square$

Now we can prove the main result on exceptional maximal subgroups. It will enable us to generalize many results that are valid for primes in $\sigma(M)$.

Theorem 11.7. We have $M_\sigma A \triangleleft M$.

Proof. Assume this is false. Let E be a complement to M_σ in M that contains A. Then $\mathrm{r}(E) = 2$. Let

$$\tau = \{ q \in \pi(E) \mid q > p \},$$

and let $K = \mathcal{O}_\tau(E)$. By Theorem 4.20, K is a Hall τ-subgroup of E and KP is a normal Hall $\tau \cup \{p\}$-subgroup of E.

By Corollary 11.6(a), $A = \Omega_1(P)$. If A centralized K, we would have $KA = K \times A$, $A = \Omega_1(\mathcal{O}_p(KP)) \lhd E$, and $M_\sigma A \lhd M_\sigma E = M$, which we have assumed false.

Thus A does not centralize K. Take a prime divisor q of $|K : C_K(A)|$, and let Q be an A-invariant Sylow q-subgroup of K. Then A does not centralize Q.

Suppose $C_Q(A) \neq 1$. Then Q is not cyclic. Consequently $\mathrm{r}(Q) = 2$. Let $B \in \mathcal{E}^2(Q)$. By Lemma 10.4(c), A and B lie in $\mathcal{E}_p{}^*(G)$ and $\mathcal{E}_q{}^*(G)$, respectively. Moreover, $q \in \pi(C_Q(A)) \subseteq \pi(C_G(A))$. Let $Q \subseteq Q^* \in \mathcal{H}_G{}^*(A; q)$. By Proposition 10.10(c), $[A, Q^*] = 1$, contrary to the fact that $[A, Q] \neq 1$. This contradiction shows that

$$C_Q(A) = 1.$$

Let $Q_0 = Z(Q)$. By Proposition 1.6(d),

$$Q_0 = [A, Q_0]C_{Q_0}(A) = [A, Q_0].$$

With A_1 and A_2 as in Corollary 11.6(c), $A = A_1 \times A_2$ and hence

$$Q_0 = [A, Q_0] = [A_1, Q_0][A_2, Q_0].$$

By Proposition 10.11(d), with $K = Q_0$ and P being A_1 and then A_2, both subgroups $[A_1, Q_0]$ and $[A_2, Q_0]$ are normal in M. Hence $Q_0 \lhd M$ and $N_G(Q) \subseteq N_G(Q_0) = M$. But Q is a Sylow q-subgroup of M and $q \notin \sigma(M)$. This contradiction completes the proof of the theorem. $\quad\square$

12. The Subgroup E

In this section we return to the study of an arbitrary maximal subgroup M of G. Let E denote a complement of M_σ in M. We will analyze the structure of E and the way E is embedded in M and in G. Of course, $E \cong M/M_\sigma$. We will see that $\mathrm{r}(E) \leq 2$, E' is nilpotent, and every Sylow subgroup of E is abelian (Lemma 12.1 and Corollary 12.10(a)). Moreover, the hypothesis of the preceding section is satisfied if $\mathrm{r}(E) = 2$ (Proposition 12.4).

We introduce the following notation:

$$\tau_1(M) = \{ p \in \sigma(M)' \mid p \notin \pi(M') \text{ and } \mathrm{r}_p(M) = 1 \}$$
$$\tau_2(M) = \{ p \in \sigma(M)' \mid \mathrm{r}_p(M) = 2 \}$$
$$\tau_3(M) = \{ p \in \sigma(M)' \mid p \in \pi(M') \text{ and } \mathrm{r}_p(M) = 1 \}$$
$$E_{12} = \text{ some Hall } \tau_1(M) \cup \tau_2(M)\text{-subgroup of } E$$
$$E_i = \text{ some Hall } \tau_i(M)\text{-subgroup of } E_{12}, \ (i = 1, 2)$$
$$E_3 = \text{ some Hall } \tau_3(M)\text{-subgroup of } E.$$

Since $\sigma(M)$ contains $\alpha(M) = \{p \in \pi(M) \mid \mathrm{r}_p(M) \geq 3\}$, $\pi(M)$ is the disjoint union of $\sigma(M)$, $\tau_1(M)$, $\tau_2(M)$, and $\tau_3(M)$. Whereas $\sigma(M)$ is not empty, by Theorem 10.2(e), each (but not all) of the sets $\tau_i(M)$ could be empty. Since E is a complement to M_σ in M, we have $E \cap M' = E'$ and hence

$$\pi(E) = \tau_1(M) \cup \tau_2(M) \cup \tau_3(M)$$
$$\tau_1(M) = \{p \in \pi(E) - \pi(E') \mid \mathrm{r}_p(E) = 1\}$$
$$\tau_2(M) = \{p \in \pi(E) \mid \mathrm{r}_p(E) = 2\}$$
$$\tau_3(M) = \{p \in \pi(E') \mid \mathrm{r}_p(E) = 1\}.$$

In the following lemma we collect some easy consequences of our definitions.

Lemma 12.1. (a) E' is nilpotent.
  (b) $E_3 \subseteq E'$ and $E_3 \lhd E$.
  (c) If $E_2 = 1$, then $E_1 \neq 1$.
  (d) E_1 and E_3 are cyclic.
  (e) $E = E_1 E_2 E_3$, $E_{12} = E_1 E_2$ and $E_2 E_3 \lhd E$ and $E_2 \lhd E_{12}$.
  (f) $C_{E_3}(E) = 1$.
  (g) If $p \in \tau_2(M)$ and $A \in \mathcal{E}_p{}^2(M)$, then $A \in \mathcal{E}_p{}^*(G)$ and $p \notin \beta(G)$.

Proof. By Theorem 10.2, M'/M_σ is nilpotent. Thus, since $M_\sigma \cap E = 1$, we know E' is nilpotent, which is (a).
  Take $p \in \tau_1(M) \cup \tau_3(M)$ and let P be a Sylow p-subgroup of E.
  Since $\mathrm{r}(P) = \mathrm{r}_p(E) = 1$ and p is odd, P is cyclic by Lemma 4.5. Thus E_1 is cyclic because $E_1' \subseteq E_1 \cap M' = 1$ by definition of $\tau_1(M)$ and E_1.
  By (a), $\mathcal{O}_{p'}(E)P \lhd E$. The Frattini argument now yields

$$(12.1) \qquad E = \mathcal{O}_{p'}(E)N_E(P) = \mathcal{O}_{p'}(E)PK,$$

where K is a complement to P in $N_E(P)$.
  If $p \in \tau_3(M)$, then $P \cap E' \neq 1$ and, by (12.1), this implies that $[P, K] \neq 1$. By Proposition 1.6(d), $P = [P, K] \times C_P(K)$. Therefore $[P, K] = P$ and $C_P(K) = 1$ because P is cyclic.
  This proves that $E_3 \subseteq E'$. Since E' is nilpotent, E_3 is cyclic and is normal in E. Now we have (b) and (d). Moreover, we have shown that $C_{E_3}(E) = 1$, because $C_P(K) = 1$ in the argument above. This yields (f).
  For (e), recall that $E_3 \lhd E$ and $E_1 \cap E' = 1$, and then (c) follows from the fact that $E_3 \subseteq E' \subset E$. Finally, (g) is just Lemma 10.4(c). $\square$

Lemma 12.2. Suppose p is a prime, X is a nonidentity p-subgroup of M, and $M^* \in \mathcal{M}(N_G(X))$. Then

  (a) $p \in \sigma(M^*) \cup \tau_2(M^*)$, and
  (b) if $p \in \sigma(M)$ and $M \neq M^*$ or if $p \in \tau_1(M) \cup \tau_3(M)$, then M^* is not conjugate to M in G.

Proof. (a) Suppose that $p \notin \sigma(M^*)$. Then $\mathrm{r}_p(M^*) = 2$ by Lemma 10.5. Thus $p \in \tau_2(M)$.

(b) Suppose that M^* is conjugate to M. Then it follows that $\tau_i(M) = \tau_i(M^*)$ for $i = 1, 2, 3$. Thus, by (a), $p \in \sigma(M)$. Hence, by Theorem 10.1(b), $C_G(X)$ acts transitively on the set $\{ M^g \mid g \in G \text{ and } X \subseteq M^g \}$. This set contains M and M^*, but consists of M^* alone because $C_G(X) \subseteq N_G(X) \subseteq M^*$. In particular, $M = M^*$. $\square$

Lemma 12.3. Suppose $M^* \in \mathcal{M} - \{M\}$, p is a prime, $A \in \mathcal{E}_p{}^2(M \cap M^*)$, and $N_G(A_0) \subseteq M^*$ for some $A_0 \in \mathcal{E}^1(A)$.

(a) If $p \notin \sigma(M)$, then A centralizes $M_\sigma \cap M^*$.

(b) If $p \in \sigma(M) - \alpha(M)$, then A centralizes $M_\alpha \cap M^*$.

Proof. If $p \in \sigma(M^*)$, then $A \subseteq M^*_\sigma$. If $p \notin \sigma(M^*)$, then M^*, A, and A_0 satisfy the hypotheses on M, A, and A_0 stated at the beginning of Section 11, and it follows from Theorem 11.7 that $M^*_\sigma A \lhd M^*$.

Therefore, in either case, every A-invariant p'-subgroup K of M^* satisfies

$$[K, A] \subseteq K \cap M^*_\sigma A \subseteq K \cap M^*_\sigma.$$

If $p \in \sigma(M) - \alpha(M)$, we apply this to $K = M_\alpha \cap M^*$. In this case, M^* is not conjugate to M by Lemma 12.2(b). Then $[K, A] \subseteq M_\alpha \cap M^*_\sigma = 1$ by Lemma 10.12(a). This proves (b).

For the proof of (a), suppose that $p \notin \sigma(M)$ and $[K, A] \neq 1$, where $K = M_\sigma \cap M^*$. Then $M_\sigma \cap M^*_\sigma \neq 1$. Consequently, by Corollary 11.4, M^* does not satisfy the hypotheses of Section 11.

It follows that $p \in \sigma(M^*)$. In particular, M^* is not conjugate to M in G because $p \notin \sigma(M)$. Let S be a Sylow p-subgroup of M^*_σ that contains A. Then $M^*_\alpha S \lhd M^*$ because M^*_σ / M^*_α is nilpotent by Theorem 10.2. Thus

$$1 \subset [K, A] \subseteq K \cap M^*_\alpha S \subseteq K \cap M^*_\alpha \subseteq M_\sigma \cap M^*_\alpha,$$

contrary to Lemma 10.12(a).

This contradiction completes the proof of (a) and of the lemma. $\square$

Proposition 12.4. Suppose p is a prime and $A \in \mathcal{E}_p{}^2(M)$. Then

(a) $C_G(A) \subseteq M$, and

(b) if $\mathcal{M}(N_G(A_0)) \neq \{M\}$ for every $A_0 \in \mathcal{E}^1(A)$, then $p \in \sigma(M)$, $M_\alpha = 1$, and M_σ is nilpotent.

Proof. We can assume that the hypothesis of (b) holds since otherwise (a) holds. Take any subgroup $X \in \mathcal{E}^1(A)$. Then there exists a subgroup $M^* \in \mathcal{M}(N_G(X)) - \{M\}$. By the Uniqueness Theorem,

$$\mathrm{r}(C_M(A)) \leq \mathrm{r}(C_M(X)) \leq 2.$$

Assume first that $p \notin \sigma(M)$. Then, by Lemma 12.3(a),

$$C_{M_\sigma}(X) \subseteq M_\sigma \cap M^* \subseteq C_M(A).$$

Thus, by Proposition 1.16,

$$M_\sigma = \left\langle C_{M_\sigma}(Y) \mid Y \in \mathcal{E}^1(A) \right\rangle \subseteq C_M(A),$$

contrary to Proposition 10.11(b). This proves that $p \in \sigma(M)$. Let P be a Sylow p-subgroup of M_σ that contains A. Then $Z = \Omega_1(Z(P)) \subseteq A$ because $AZ \in \mathcal{E}_p(C_M(A))$ and $\mathrm{r}(C_M(A)) \leq 2$.

Note that $Z \neq 1$ because $P \neq 1$. Thus we can choose our subgroup X in Z and get

$$\mathrm{r}(P) \leq \mathrm{r}(C_M(X)) \leq 2.$$

Thus $p \in \sigma(M) - \alpha(M)$. Now, using Lemma 12.3(b), we see that

$$M_\alpha = \left\langle C_{M_\alpha}(Y) \mid Y \in \mathcal{E}^1(A) \right\rangle \subseteq C_M(A).$$

It follows that $M_\alpha = 1$ because $\mathrm{r}(C_M(A)) \leq 2$, and then $M_\sigma \cong M_\sigma/M_\alpha$ is nilpotent by Theorem 10.2. Therefore $P \lhd M$ and

$$C_G(A) \subseteq C_G(Z) \subseteq N_G(Z) = M,$$

which completes the proof of the proposition. □

Now we are in a position to obtain substantial information about the unpleasant case when $\tau_2(M)$ is not empty.

Theorem 12.5. Suppose that $\tau_2(M)$ is not empty. Let $p \in \tau_2(M)$ and $A \in \mathcal{E}_p{}^2(M)$. Then

(a) M_σ is nilpotent,
(b) M has abelian Sylow p-subgroups and every Sylow p-subgroup P of M such that $A \subseteq P$ satisfies $\Omega_1(P) = A$ and $N_G(P) \not\subseteq M$,
(c) $M_\sigma A \lhd M$,
(d) $C_{M_\sigma}(A) = 1$,
(e) $M_\sigma \cap M^* = 1$ for every $M^* \in \mathscr{M}(A) - \{M\}$, and
(f) there exists $A_1 \in \mathcal{E}^1(A)$ such that $C_{M_\sigma}(A_1) = 1$.

Proof. By Proposition 12.4, $N_G(A_0) \subseteq M$ for some subgroup $A_0 \in \mathcal{E}^1(A)$ because $p \notin \sigma(M)$. Thus we have Hypothesis 11.1, and everything except (e) follows from Theorems 11.3, 11.5, 11.7, and Corollary 11.6.

Suppose that $M^* \in \mathscr{M}(A) - \{M\}$.

If $N_G(A_0) \subseteq M^*$ for some $A_0 \in \mathcal{E}^1(A)$, then (d) and Lemma 12.3(a) yield

$$M_\sigma \cap M^* \subseteq C_{M_\sigma}(A) = 1.$$

Otherwise, Proposition 12.4(b) shows that $A \subseteq \mathcal{O}_p(M^*)$, and we reach the same conclusion because

$$[M_\sigma \cap M^*, A] \subseteq M_\sigma \cap \mathcal{O}_p(M^*) = 1. □$$

Recall our choice of E, E_1, E_2, and E_3 at the beginning of this section.

Corollary 12.6. Suppose that $\tau_2(M)$ is not empty. Let $p \in \tau_2(M)$ and $A \in \mathcal{E}_p{}^2(E)$. Then

(a) $A \lhd E$ and $\mathcal{E}_p{}^1(E) = \mathcal{E}^1(A)$,

(b) $C_G(A) \subseteq N_M(A) = E$ and $N_G(A) \not\subseteq M$,

(c) $\mathcal{M}(C_G(X)) = \{M\}$ for each $X \in \mathcal{E}^1(A) = \mathcal{E}_p{}^1(E)$ such that $C_{M_\sigma}(X) \neq 1$,

(d) $C_{M_3}(x) = 1$ for each $x \in E_3{}^\#$,

(e) $C_{M_\sigma}(x) = 1$ for each $x \in C_{E_1}(A)^\#$, and

(f) if $M^* \in \mathcal{M}$ is not conjugate to M, then $M_\sigma \cap M_\sigma^* = 1$ and $\sigma(M)$ is disjoint from $\sigma(M^*)$.

Proof. Recall that $M = M_\sigma E$ and $M_\sigma \cap E = 1$. Thus $A \lhd E$ because $M_\sigma A \lhd M$ by Theorem 12.5(c). Hence A is contained in every Sylow p-subgroup P of E and, for each such P, Theorem 12.5(b) asserts that

(12.2) $\Omega_1(P) = A$ and $N_G(P) \not\subseteq M$.

This proves (a) and shows that $N_g(P) \subseteq N_G(A)$ and

$$C_M(A) \subseteq N_M(A) = N_M(A) \cap M_\sigma E = (N_M(A) \cap M_\sigma)E = C_{M_\sigma}(A)E.$$

By Theorem 12.5(d), $C_{M_\sigma}(A) = 1$. Therefore, by (12.2), we obtain (b).

By Theorem 12.5(e), $M_\sigma \cap M^* = 1$ for every $M^* \in \mathcal{M}(A) - \{M\}$. This yields (c) immediately. It also yields (d) and (e) because we can assume x has prime order and apply Lemma 12.2(b) to deduce that $N_G(\langle x \rangle) \not\subseteq M$. Since M_σ is nilpotent, by Theorem 12.5 (a), Lemma 10.12(b) yields (f). $\quad\square$

Theorem 12.7. Suppose that $p \in \tau_2(M)$, $A \in \mathcal{E}_p{}^2(E)$, and assume that G has nonabelian Sylow p-subgroups. Then

(a) $\tau_2(M) = \{p\}$,

(b) $A_0 = C_A(M_\sigma)$ has order p and satisfies $F(M) = M_\sigma \times A_0$,

(c) every $X \in \mathcal{E}_p{}^1(E) - \{A_0\}$ satisfies $C_{M_\sigma}(X) = 1$ and $C_G(X) \not\subseteq M$,

(d) A_0 has a complement E_0 in E, and

(e) $\pi(C_{E_0}(x)) \subseteq \tau_1(M)$ for every $x \in M_\sigma{}^\#$.

Proof. Choose Sylow p-subgroups P of E and S of G such that

$$A \subseteq P \subseteq S.$$

Since we have P abelian and $C_G(A) \subseteq E$ by Theorem 12.5(b) and Corollary 12.6(b),

$$C_S(A) = P \subset S.$$

Suppose that $q \in \tau_2(M) - \{p\}$. Take $B \in \mathcal{E}_q{}^2(E)$. Then A centralizes B because, by Corollary 12.6(a), both A and B are normal in E. Furthermore, $C_G(B) \subseteq E$ by Corollary 12.6(b), and both A and B lie in $\mathcal{E}^*(G)$ by Lemma 12.1(g). Therefore, by Proposition 10.10(c), $C_G(B)$ contains a

Sylow p-subgroup of G, contrary to the fact that $P \subset S$ and $C_G(B) \subseteq E$. This proves (a).

By Proposition 1.16,

$$M_\sigma = \langle\, C_{M_\sigma}(X) \mid X \in \mathcal{E}_p{}^1(A)\,\rangle,$$

and therefore there exists a subgroup $A_0 \in \mathcal{E}^1(A)$ such that $C_{M_\sigma}(A_0) \neq 1$. By Corollary 12.6(c), this implies that

$$\mathcal{M}(C_G(A_0)) = \{M\}.$$

We know that $S \nsubseteq M$ and $A_0 \nsubseteq Z(S)$ because P is a Sylow p-subgroup of M.

Suppose that $X \in \mathcal{E}_p{}^1(E) - \{A_0\}$. By Corollary 12.6(a), $X \subseteq A$. If $X \subseteq Z(S)$, then $C_G(X) \nsubseteq M$ and hence $C_{M_\sigma}(X) = 1$. If $X \nsubseteq Z(S)$, then, by Lemma 10.13(c), $X = A_0{}^g$ for some $g \in S$. Then g lies outside P and hence outside M because P is abelian. Then $\mathcal{M}(C_G(X)) = \{M^g\} \neq \{M\}$, and therefore $C_{M_\sigma}(X) = 1$ and $C_G(X) \nsubseteq M$. Now

$$M_\sigma = \langle\, C_{M_\sigma}(X) \mid X \in \mathcal{E}_p{}^1(A)\,\rangle = C_{M_\sigma}(A_0),$$

and the proof of (c) is complete.

By Lemma 10.13(b),

$$P = C_S(A) = A_0 \times Z$$

for some cyclic subgroup Z. By (c), $C_Z(M_\sigma) = 1$ and hence $C_P(M_\sigma) = A_0$. Thus A_0 is a Sylow p-subgroup of $C_M(M_\sigma)$. Moreover, $A_0 = C_A(M_\sigma) \lhd M$ because $A \lhd E$ and $M = M_\sigma E$. By (a) and Lemma 12.2(a),

$$\pi(F(M)) \subseteq \sigma(M) \cup \tau_2(M) = \sigma(M) \cup \{p\}.$$

This shows that $F(M) = M_\sigma \times A_0$ because M_σ is nilpotent. Now we have (b).

Next we prove (d); then (e) follows from (a), (c), and Corollary 12.6(d).

By (a), P is a Hall $\tau_2(M)$-subgroup of E, say $P = E_2$. Recall that, by Lemma 12.1, $E_3 \lhd E$, $E_2 E_3 \lhd E$, $E = E_1 E_2 E_3$, and $E_2 \lhd E_{12} = E_1 E_2$.

Clearly $Y = \{\, x^p \mid x \in P \,\}$ is the subgroup of index p in Z and is invariant under $N_E(P)$ and hence under E_1. Thus E_1 acts on the elementary abelian p-group P/Y, and Maschke's Theorem (or Theorem 1.6(d)) yields an E_1-invariant complement P_0/Y to $A_0 Y/Y$ in P/Y. Therefore $P = A_0 \times P_0$, and $E_0 = E_1 P_0 E_3$ is a complement of A_0 in E. This completes the proof of (d) and of the theorem. $\square$

Our next result is, in a sense, a complement to the theorem just proved.

Lemma 12.8. Suppose that $p \in \tau_2(M)$, $A \in \mathcal{E}_p{}^2(E)$, and S is a Sylow p-subgroup of G that contains A. Assume that S is abelian. Then

(a) E_2 is an abelian normal subgroup of E,
(b) E_2 is a Hall $\tau_2(M)$-subgroup of G,
(c) $S \subseteq N_G(S)' \subseteq F(E) \subseteq C_G(S) \subseteq E$,
(d) $N_G(A) = N_G(S) = N_G(E_2) = N_G(E_2 E_3) = N_G(F(E))$,
(e) every $X \in \mathcal{E}^1(E_1)$ for which $C_{M_\sigma}(X) = 1$ lies in $Z(E)$, and
(f) for every subgroup X of $N_G(S)$, we have $C_S(X) \lhd N_G(S)$ and $[S, X] \lhd N_G(S)$.

Proof. By Theorem 12.7(a), each $p \in \tau_2(M)$ satisfies our assumption that G has abelian Sylow p-subgroups. Since

$$S \subseteq C_G(A) \subseteq E,$$

it follows that E_2 is a Hall subgroup of G, which proves (b).

Since $A \lhd E$, $C_G(A) \subseteq E$, and S is abelian,

$$F(N_G(A)) = F(C_G(A)) = F(E).$$

In particular, $r(F(N_G(A))) \leq 2$. Therefore, by Theorem 4.20(a),

$$N_G(A)' \subseteq F(N_G(A)).$$

Moreover, $N_G(S) \subseteq N_G(A)$ because $A = \Omega_1(S)$ and, by Corollary 10.7(a),

$$S \subseteq N_G(S)'.$$

This yields (c) and shows that $S \lhd E$. By our remark at the beginning of the proof, E_2 is abelian, and it follows that $E_2 \lhd E$. Now (a) holds and each subgroup in the series

$$A \subseteq S \subseteq E_2 \subseteq E_2 E_3 \subseteq F(E)$$

is characteristic in its successor. Therefore, because $N_G(A) \subseteq N_G(F(E))$, (d) holds.

For (f), note that $C_G(S)X \lhd N_G(S)$ because $N_G(S)' \subseteq C_G(S)$. Clearly $C_S(X) = C_S(C_G(S)X) \lhd N_G(S)$ and similarly $[S, X] \lhd N_G(S)$.

Finally, let $K = E_2 E_3 = E_2 \times E_3$ for the proof of (e). By (c) and (d), $N_G(S)' \subseteq C_G(K)$ and $K \lhd N_G(S)$. Hence, as in (f), $[K, X] \lhd N_G(S)$. Therefore, because $p \notin \sigma(M)$, we know that $N_G([K, X]) \not\subseteq M$. However $[K, X] \lhd M$ by Proposition 10.11(d). Therefore $[K, X] = 1$, and hence $E = KE_1 \subseteq C_E(X)$ because E_1 is cyclic. $\square$

Corollary 12.9. Suppose $p \in \tau_2(M)$, $A \in \mathcal{E}_p{}^2(E)$, $q \in \tau_1(M)$, $Q \in \mathcal{E}_q{}^1(E)$, $C_{M_\sigma}(Q) = 1$, and $[A, Q] \neq 1$. Let $A_0 = [A, Q]$ and $A_1 = C_A(Q)$. Then

(a) $A_0 \in \mathcal{E}^1(A)$ and $A_0 = C_A(M_\sigma) \lhd M$,

(b) A_0 is not conjugate to A_1 in G, and

(c) $A_1 \in \mathcal{E}^1(A)$ and $C_G(A_1) \not\subseteq M$.

Proof. Since $A \lhd E$, we have $A = A_0 \times A_1$ and Proposition 10.11(d) yields (a). Thus $A_1 \in \mathcal{E}^1(A)$ and $N_G(A_0) = M$. Since $\mathrm{r}_q(M) = 1$ and $Q \not\subseteq C_M(A_0) = C_G(A_0)$, we know that $C_G(A_0)$ is a q'-group. In particular, (b) holds.

By Lemma 12.8(e) for $X = Q$, we are in the situation of Theorem 12.7, and Theorem 12.7(c) yields (c). $\square$

Corollary 12.10. (a) Every nilpotent $\sigma(M)'$-subgroup of M is abelian.

(b) The groups E_2 and E' are abelian.

(c) Suppose $p \in \tau_2(M)$ and $A \in \mathcal{E}_p{}^2(E)$. Then $E_2 E_3 \subseteq C_E(A) \lhd E$ and $\pi(E/C_E(A)) \subseteq \tau_1(M)$.

(d) Suppose $p \in \sigma(M)$ and P is a noncyclic p-subgroup of M. Then $N_G(P) \subseteq M$.

(e) Suppose $x \in M^\#$, $\pi(\langle x \rangle) \subseteq \tau_2(M)$, and $C_{M_\sigma}(x) \neq 1$. Then $\mathcal{M}(C_G(x)) = \{M\}$.

Proof. By Lemma 12.1(d) and Theorem 12.5(b), M has abelian Sylow p-subgroups for every prime p in $\tau_1(M) \cup \tau_2(M) \cup \tau_3(M)$. Thus (a) holds. Moreover, E' is abelian because E' is nilpotent by Lemma 12.1(a). Furthermore, Theorems 12.5(b), 12.7(a), and Lemma 12.8(a) show that E_2 is abelian. Thus Theorem 12.5(e) yields (e) because we can assume that x lies in E_2. For (c), recall that A, $E_3 \lhd E$ and $E = E_1 E_2 E_3$. Finally, in the situation of (d), there exists $A \in \mathcal{E}_p{}^2(P)$ because p is odd and, by Theorem 10.1(c) and Proposition 12.4(a),

$$N_G(P) = N_M(P)C_G(P) \text{ and } C_G(P) \subseteq C_G(A) \subseteq M. \quad \square$$

Lemma 12.11. Suppose $p \in \tau_2(M)$, $A \in \mathcal{E}_p{}^2(E)$, and $M^* \in \mathcal{M}(N_G(A))$. Then

(a) $\tau_2(M) \subseteq \sigma(M^*) - \beta(M^*)$,

(b) $\pi(E/C_E(A)) \subseteq \tau_1(M^*) \cup \tau_2(M^*)$, and

(c) if $q \in \pi(E/C_E(A)) \cap \pi(C_E(A))$, then $q \in \tau_2(M^*)$, some Sylow p-subgroup of G is normal in M^*, and M^* contains an abelian Sylow q-subgroup of G.

Proof. Since $N_G(A) \subseteq M^*$, Corollary 12.6(b) shows that $p \notin \tau_2(M^*)$. Thus p lies in $\sigma(M^*)$. Take $p^* \in \tau_2(M)$ and $A^* \in \mathcal{E}_{p^*}{}^2(E)$. If $p^* = p$, choose $A^* = A$. By Corollary 12.6(a), both A and A^* are normal in E. Consequently $A^* \subseteq M^*$ and $1 \subset A \subseteq C_{M_\sigma^*}(A^*)$. Therefore, by Theorem 12.5(d),

$p^* \notin \tau_2(M^*)$. Thus p^* lies in $\sigma(M^*)$ but not in $\beta(M^*)$ because $p^* \notin \beta(G)$ by Lemma 12.1(g). This proves (a).

Suppose that $q \in \pi(E/C_E(A))$ and Q is a Sylow q-subgroup of E. Then $Q \subseteq M^*$ and $1 \subset [A, Q] \subseteq A$ because $A \triangleleft E$. By Theorem 10.2(c) and Lemma 12.1(b), $M^{*\prime}$ contains Hall $\sigma(M^*)$- and $\tau_3(M^*)$-subgroups of M^*. Thus, if (b) is false for q, then Q lies in $M^{*\prime}$. It follows that $[A, Q] \subseteq M^*_\beta$ because $A \subseteq M^*_\sigma \subseteq M^{*\prime}$, and $M^{*\prime}/M^*_\beta$ is nilpotent by Lemma 10.8. But this contradicts the fact that $1 \subset [A, Q] \subseteq A$ and $p \notin \beta(M^*)$.

It remains to prove (c), so assume that $Q_0 \in \mathcal{E}_q^{\,1}(Q)$ and centralizes A. By Corollary 12.10(c), $q \in \tau_1(M)$. By Corollary 12.6(b), $C_G(A) \subseteq E$. Thus Q and a Sylow q-subgroup of $C_G(A)$ are cyclic, $Q_0 = \Omega_1(Q)$, and the Frattini argument yields

$$N_G(A) = C_G(A)N_{N_G(A)}(Q_0)$$

because $Q_0 \subseteq C_G(A) \triangleleft E$. Thus $r_q(C_G(A)) = 1$.

Take $M^{**} \in \mathcal{M}(N_G(Q_0))$. Then $N_G(A) \subseteq M^{**}$ because $C_G(A) \subseteq M^{**}$ by Proposition 12.4(a). By (b) and Lemma 12.2(a), both applied to M^{**} in place of M^*, q lies in $\sigma(M^{**}) \cup \tau_2(M^{**})$ and in $\tau_1(M^{**}) \cup \tau_2(M^{**})$.

Therefore $q \in \tau_2(M^{**})$ and, since $p \in \sigma(M^{**}) \cap \sigma(M^{**})$ by (a), it follows from Corollary 12.6(f) that M^{**} is conjugate to M^* in G. Now Theorem 10.1(b) shows that $M^* = M^{**}$ because $A \subseteq M^* \cap M^{**}$ and $C_G(A) \subseteq M^*$.

Since $q \in \tau_2(M^*)$, it follows from Theorem 12.5(a) that M^*_σ is nilpotent. Hence $\mathcal{O}_p(M^*)$ is a Sylow p-subgroup of G. Since $Q_0 = \Omega_1(Q) \subset Q \subseteq M^*$, it follows that Q_0 does not have a complement in M^*. Consequently, by Theorem 12.7(d), G has abelian Sylow q-subgroups and one of them lies in $N_G(Q_0) = N_{M^*}(Q_0)$. $\square$

Theorem 12.12. Suppose $C_{M_\sigma}(e) = 1$ for each $(\tau_1(M) \cup \tau_3(M))$-element $e \in E^\#$. Then

(a) E contains an abelian normal subgroup A_0 such that $C_E(x) \subseteq A_0$ for every $x \in M_\sigma^\#$, and

(b) E contains a subgroup E_0 of the same exponent as E such that $E_0 M_\sigma$ is a Frobenius group with Frobenius kernel M_σ.

Proof. If $E = E_1 E_3$, then (a) and (b) hold with $A_0 = 1$ and $E_0 = E$. Hence we can assume that $\tau_2(M)$ is not empty.

If G has nonabelian Sylow p-subgroups, then Theorem 12.7 provides subgroups A_0 and E_0 as required. Therefore we can assume the hypotheses, notation, and conclusions of the complementary result Lemma 12.8.

Obviously $A_0 = E_2$ satisfies (a) and, to get (b), it suffices to find a cyclic normal subgroup $Z = Z_p$ of E having the same exponent as S and satisfying $C_{M_\sigma}(z) = 1$ for every $z \in Z^\#$, i.e., $C_{M_\sigma}(\Omega_1(Z)) = 1$, for then we can let E_0 be the product of $E_1 E_3$ and all these Z_p, one for each $p \in \tau_2(M)$.

Since $p \notin \sigma(M)$, it follows that $N_G(S) \nsubseteq M$. Thus every $N_G(S)$-invariant nonidentity cyclic subgroup Z of S satisfies $C_{M_\sigma}(\Omega_1(Z)) = 1$ because otherwise, by Corollary 12.6(c),

$$N_G(S) \subseteq N_G(\Omega_1(Z)) \subseteq M.$$

Assume first that $C_E(S) = E$. Since S is abelian of rank two,

$$S = Y \times Z,$$

for some cyclic subgroups Y and Z such that

$$|Y| \leq |Z|.$$

If $|Y| < |Z|$, then $\Omega_1(Z)$ is characteristic in S, and the discussion above shows that Z is as required. If $|Y| = |Z|$, then Y and Z can be chosen in such a way that $\Omega_1(Z)$ is equal to any given $A_1 \in \mathcal{E}^1(A)$ and, by Theorem 12.5(f), at least one such A_1 satisfies $C_{M_\sigma}(A_1) = 1$.

Therefore, for the remainder of the proof, we can assume that $C_E(S) \neq E$. Take $q \in \pi(E/C_E(S))$ and let Q be a Sylow q-subgroup of $N_G(S)$ that contains a Sylow q-subgroup Q_1 of E. Since $C_S(Q_1) \subset S$, it follows from Proposition 1.6(e) that $Q_1 \nsubseteq C_E(A)$. Therefore, by Corollary 12.10(c), $q \in \tau_1(M)$ and Q_1 is cyclic. Since $C_G(S) \subseteq E$,

$$Q_0 = C_Q(S) \subset Q_1.$$

Suppose Q/Q_0 acts regularly on S. Then Proposition 3.9 shows that Q/Q_0 is cyclic. Hence $\Omega_1(Q/Q_0) \subseteq Q_1/Q_0$ and $\Omega_1(Q) \subseteq Q_1$. Therefore $r_q(N_G(S)) = r(Q) = 1$ because Q_1 is cyclic. However, by Lemma 12.8(e), $\Omega_1(Q_1)$ centralizes A, and therefore Lemma 12.11(c) yields $r_q(N_G(S)) = 2$. This contradiction shows that

$$1 \subset C_S(X) \subset S$$

for some subgroup X of Q.

Now $S_0 = C_S(X)$ and $S_1 = [S, X]$ are cyclic because

$$S = S_0 \times S_1.$$

Moreover, by Lemma 12.8(f), both S_0 and S_1 are normal in $N_G(S)$ and hence act regularly on M_σ. Now define $Z = S_0$ if $|S_0| \geq |S_1|$ and $Z = S_1$ if $|S_0| < |S_1|$. Then Z is as required, and the proof of Theorem 12.12 is complete. $\square$

Now we leave $\tau_2(M)$ and turn to $\sigma(M)$. First we derive an important uniqueness result from our basic Proposition 12.4.

Theorem 12.13. Every nonabelian p-subgroup of G (for every prime p) lies in $\mathcal{U}$.

Proof. Suppose $p \in \pi(G)$ and P is a nonabelian P-subgroup of G that lies in two distinct maximal subgroups M and M^* of G. Then

$$p \in \sigma(M) \cap \sigma(M^*) \text{ and } N_G(P) \subseteq M \cap M^*$$

by Corollary 12.10(a) and (d).

We can assume that P is a Sylow p-subgroup of $M \cap M^*$. Then P is a Sylow p-subgroup of M and of M^* and hence of G. By the Uniqueness Theorem, $\mathrm{r}(P) \le 2$. Therefore, since P is not cyclic and p is odd, we know that $\mathrm{r}(P) = 2$ and, by Corollary 10.7(b), there exists a nonabelian subgroup $Q \subseteq P$ of order p^3 and exponent p.

Now $Z = Z(Q) = Q'$ has order p, Q/Z acts on $K = C_{M_\alpha}(Z) = N_{M_\alpha}(Z)$, and Proposition 1.16 yields

$$K = \left\langle \, C_K(\overline{A}) \mid \overline{A} \in \mathcal{E}^1(Q/Z) \, \right\rangle.$$

Clearly, if $A/Z \in \mathcal{E}^1(Q/Z)$, then $A \in \mathcal{E}^2(Q)$ and, furthermore, we have $C_K(A/Z) = C_K(A) \subseteq M^*$ by Proposition 12.4(a). Thus $K \subseteq M^*$ and, since

$$M = (M \cap M^*)M_\alpha$$

by Corollary 10.9(b), it follows from Lemma 6.5(b) that $N_M(Z) \subseteq M^*$. In particular, $\mathcal{M}(N_G(Z)) \ne \{M\}$. Furthermore, $M_\alpha \ne 1$.

Thus, for any $A \in \mathcal{E}^2(Q)$, Proposition 12.4(b) provides a subgroup $A_0 \in \mathcal{E}^1(A) - \{Z\}$ such that $\mathcal{M}(N_G(A_0)) = \{M\}$. Similarly, there exists a subgroup $A_0^* \in \mathcal{E}^1(A) - \{Z\}$ which satisfies $\mathcal{M}(N_G(A_0^*)) = \{M^*\}$. Now A_0^* is not conjugate to A_0 in $M \cap M^*$ and hence not in Q, contrary to the structure of Q. This contradiction completes the proof of Theorem 12.13. $\square$

Corollary 12.14. Suppose $p \in \sigma(M)$, $X \in \mathcal{E}_p{}^1(M)$, and P is a Sylow p-subgroup of M_σ. Assume $p \in \beta(M)$ or $X \subseteq M_\sigma{}'$. Then

$$\mathcal{M}(C_G(X)) = \mathcal{M}(P) = \{M\}.$$

Proof. We can assume that $X \subseteq P$ and, by the Uniqueness Theorem, that $\mathrm{r}(C_P(X)) \le 2$. By Corollary 5.4 and the definition of a narrow p-group (Section 5), P is narrow if $\mathrm{r}(P) \ge 3$. Thus $p \notin \beta(G)$.

By Lemma 10.8(c), $M_\sigma = PO_{p'}(M_\sigma)$. Therefore

$$1 \subset X \subseteq P \cap M_\sigma{}' = P'.$$

Since $\mathrm{r}(C_P(X)) \le 2$, Theorem 5.3(d) shows $\mathrm{r}(P) \le 2$. Then, by Corollary 10.7(b),

$$X \subseteq P' \subseteq Z(P).$$

Hence $P \subseteq C_M(X)$. Since P is not abelian, $P \in \mathcal{U}$ by Theorem 12.13. This concludes the proof of the corollary. $\square$

The next result gives some control over the embedding of M_σ in G. We shall see later (Theorem 13.9) that case (d) cannot occur.

Proposition 12.15. Suppose $q \in \sigma(M)$, X is a nonidentity q-subgroup of M, and $M^* \in \mathcal{M}(N_G(X)) - \{M\}$. Let S be a Sylow q-subgroup of $M \cap M^*$ that contains X. Then S, M, and M^* satisfy the following conditions.

   (a) M^* is not conjugate to M in G.
   (b) $N_G(S) \subseteq M$.
   (c) S is a Sylow q-subgroup of M^*.
   (d) If $q \in \sigma(M^*)$, then
        (1) $M^* = (M \cap M^*)M^*_\beta$,
        (2) $\tau_1(M^*) \subseteq \tau_1(M) \cup \alpha(M)$, and
        (3) $M_\beta = M_\alpha \neq 1$.
   (e) If $q \notin \sigma(M^*)$, then
        (1) $q \in \tau_2(M^*)$,
        (2) $\pi(M) \cap \sigma(M^*) \subseteq \beta(M^*)$, and
        (3) $M \cap M^*$ is a complement to M^*_σ in M^*.

Proof. Let T be a Sylow q-subgroup of M that contains S. If S is not cyclic, then, by Corollary 12.10(d), $N_G(S) \subseteq M$. If S is cyclic, then $N_G(S) \subseteq N_G(X) \subseteq M^*$, $S \subseteq N_T(S) \subseteq M \cap M^*$, and $S = N_T(S)$. Therefore $S = T$, and $N_G(S) \subseteq M$ because $p \in \sigma(M)$. This proves (b), and (a) follows from Lemma 12.2(b). Clearly (b) implies (c). Now

$$N_G(S) \subseteq M^* \text{ if and only if } q \in \sigma(M^*).$$

By Lemma 12.2(a), $q \in \sigma(M^*) \cup \tau_2(M^*)$.

Assume first that $q \in \tau_2(M^*)$. Take $A \in \mathcal{E}^2(S)$ and let E^* be a complement of M^*_σ in M^* that contains A. Then, by Theorem 12.5(e) and Corollary 12.6(a), $M^*_\sigma \cap M = 1$ and $A \triangleleft E^*$. By Corollary 12.10(d), $E^* \subseteq N_G(A) \subseteq M$. Thus $M \cap M^* = M \cap M^*_\sigma E^* = (M \cap M^*_\sigma)E^* = E^*$.

Suppose that p is a prime in $\pi(M) \cap \sigma(M^*)$ not lying in $\beta(M^*)$. Then

$$p \in \sigma(M^*) - \beta(M^*) \text{ and } q \in \pi(M^*) - \sigma(M^*).$$

Furthermore, since $q \notin \beta(G)$ by Lemma 12.1(g), and $\sigma(M)$ is disjoint from $\sigma(M^*)$ by Corollary 12.6(f),

$$q \in \sigma(M) - \beta(M) \text{ and } p \in \pi(M) - \sigma(M).$$

As $p \in \sigma(M^*)$ and $C_G(A) \subseteq E^*$ by Corollary 12.6(b), the group $C_G(A)$ is a p'-group. Therefore A does not centralize a Sylow p-subgroup of M^*_σ and no Sylow p-subgroup of M centralizes a Sylow q-subgroup of M_σ. By Corollary 10.9(a), $p > q$ and $q > p$. This contradiction completes the proof of (e).

It remains to discuss the case when $q \in \sigma(M^*)$. Here $N_G(S) \subseteq M^*$ and S is a Sylow q-subgroup of G. By Corollary 10.9(b),

$$M = (M \cap M^*)M_\alpha \text{ and } \alpha(M) = \beta(M)$$

and similarly for M^*. Thus M_α, $M^*_\alpha \neq 1$.

Suppose that $r \in \tau_1(M^*)$ and R is a Sylow r-subgroup of $M \cap M^*$. Then R is a Sylow r-subgroup of $M^* = (M \cap M^*)M^*_\alpha$ and has a normal complement in M^* and in $M \cap M^*$. Thus, if $r \notin \alpha(M)$, the same holds for $M = (M \cap M^*)M_\alpha$.

This proves that $\tau_1(M^*) \subseteq \tau_1(M) \cup \alpha(M)$. $\square$

Corollary 12.16. Suppose Y is a nonidentity $\sigma(M)$-subgroup of G. Then Y is conjugate to a subgroup of M_σ and for every $p \in \pi(E) \cap \beta(G)'$ and every $H \in \mathcal{M}(Y)$ not conjugate to M,

(a) $r_p(N_H(Y)) \leq 1$, and
(b) if $p \in \tau_1(M)$, then $p \notin \pi(N_H(Y)')$.

Proof. Since Y is solvable, Y must have a nonidentity characteristic q-subgroup X for some prime q. Then $q \in \sigma(M)$ and M_σ contains a Sylow q-subgroup of G. Replacing Y by some conjugate if necessary, we can therefore assume that $X \subseteq M_\sigma$.

As part of the proof of (a), note that if M contains some $A \in \mathcal{E}_p{}^2(H)$, then $p \in \tau_2(M)$, and $M_\sigma \cap H = 1$ by Theorem 12.5(e), contrary to the fact that $1 \subset X \subseteq M_\sigma \cap H$.

So, if $N_G(X) \subseteq M$, then everything is clear because

$$N_G(Y) \subseteq N_G(X) \subseteq M \text{ and hence } (N_G(Y))' \subseteq M'.$$

Hence we can assume that $N_G(X) \nsubseteq M$. Take $M^* \in \mathcal{M}(N_G(X))$. By Proposition 12.15(a) and (e), we know that $q \in \sigma(M^*) \cup \tau_2(M^*)$ and

$$(12.3) \qquad \begin{array}{l} M^* \text{ is not conjugate to } M \text{ in } G \text{ and} \\ \text{if } q \in \tau_2(M^*), \text{ then } \pi(M) \cap \sigma(M^*) \subseteq \beta(M^*). \end{array}$$

We define $K = M^*_\beta$ or $K = M^*_\sigma$ according as $q \in \sigma(M^*)$ or $q \in \tau_2(M^*)$. By Lemma 10.12(a) and Corollary 12.6(f), K is a $\sigma(M)'$-group. By Proposition 12.15,

$$(12.4) \qquad M^* = (M \cap M^*)K.$$

It follows that Y is conjugate to a subgroup of $M \cap M^*$ in M^*. Thus, since M/M_σ is a $\sigma(M)'$-group, Y is conjugate to a subgroup of M_σ.

To complete the proof, note first that K is a p'-group by (12.3) because $p \notin \beta(G)$. Furthermore, we can assume that $H = M^*$ because M^* contains $N_G(Y)$ and is not conjugate to M in G. Then, by (12.4), $H = (M \cap H)K$, so the very first argument gives (a). If $p \in \tau_1(M)$, then $p \notin \pi(M')$ and $(M \cap H)'K$ is a normal p'-subgroup of $H = (M \cap H)K$ that contains H'. Thus (b) follows. $\square$

Lemma 12.17. We have $C_{M_\sigma}(E) \subseteq M_\sigma'$, $[M_\sigma, E] = M_\sigma$ and, for every $g \in G - M$, the group $M_\sigma \cap M^g$ is a cyclic $\beta(M)'$-group intersecting M_σ' trivially.

Proof. The first assertion follows from Lemma 6.3(a) because the orders of M_σ and E are relatively prime, $M_\sigma \subseteq M'$, and $M = EM_\sigma$.

Let $g \in G - M$, $p \in \pi(M_\sigma \cap M^g)$, and $X \in \mathcal{E}_p^*(M_\sigma \cap M^g)$. By Theorem 10.1(b), $C_G(X) \not\subseteq M$. Therefore, by Corollaries 12.10(d) and 12.14, $|X| = p$, $p \notin \beta(M)$, and $X \not\subseteq M_\sigma'$. In particular, this proves that $M_\sigma \cap M^g$ is abelian of rank at most one, which means that $M_\sigma \cap M^g$ is cyclic. $\square$

Lemma 12.18. Suppose $p \in \tau_1(M)$, $P \in \mathcal{E}_p^1(M)$, $q \in p'$, and Q is a nonidentity P-invariant q-subgroup of M such that $C_Q(P) = 1$ and $\mathcal{M}(N_G(Q)) \neq \{M\}$.

    (a) If $M_\alpha \neq 1$ and $q \notin \alpha(M)$, then $C_{M_\alpha}(P) \neq 1$ and $C_{M_\alpha}(PQ) = 1$.

    (b) If Q is a Sylow q-subgroup of M, then $\alpha(M) = \beta(M)$ and we have the situation of (a).

Proof. (a) As $\mathcal{M}(N_G(Q)) \neq \{M\}$ by Lemma 12.2(a),

(12.5) $$r(C_{M_\alpha}(Q)) \leq 1.$$

Similarly, since $\mathcal{M}(N_G(P)) \neq \{M\}$ by Lemma 12.2(a),

(12.6) $$r(C_{M_\alpha}(P)) \leq 1.$$

We have assumed that $M_\alpha \neq 1$ and $q \notin \alpha(M)$. Take $r \in \alpha(M)$ and let R be a PQ-invariant Sylow r-subgroup of M_α. Then Q does not centralize R because $r(R) \geq 3$. Thus Theorem 1.13 yields a characteristic subgroup R_1 of R that has exponent r and is not centralized by Q. Let $R_0 = C_{R_1}(Q)$ and $N = N_{R_1}(R_0)$. Then QR_1 is not nilpotent and neither is QN/R_0. If $C_{R_1}(P) = 1$, then $C_{QR_1}(P) = 1$, and QR_1 would be nilpotent by Theorem 3.7, a contradiction. Hence, by (12.6), since R_1 has exponent r,

(12.7) $$C_{R_1}(P) \text{ has order } r \text{ and is equal to } \Omega_1(C_R(P)).$$

Suppose $C_{M_\alpha}(PQ) \neq 1$. Then we can choose r and R in such a way that $C_R(PQ) \neq 1$. By (12.7) and (12.5),

$$C_{R_1}(P) = \Omega_1(C_R(Q)) = C_{R_1}(Q) = R_0.$$

Therefore $C_{QN/R_0}(P) = 1$ and so QN/R_0 is nilpotent by Theorem 3.7, a contradiction.

    (b) We have $Q \subseteq M'$ because $C_Q(P) = 1$. As $\mathcal{M}(N_G(Q)) \neq \{M\}$, $Q \not\trianglelefteq M$. Therefore M' is not nilpotent and $M_\alpha \neq 1$ by Theorem 10.2(d). Moreover, by the Uniqueness Theorem (Theorem 9.6), $q \notin \alpha(M)$. If there exists a prime $r \in \alpha(M) - \beta(M)$, then Corollary 10.9(a)(2) shows that $C_M(Q) \in \mathcal{U}$, which is false. $\square$

Lemma 12.19. The group E' centralizes a Hall $\beta(M)'$-subgroup of M_σ.

Proof. By Corollary 10.9(a), every Sylow subgroup of E' centralizes a Hall $\beta(M)'$-subgroup of M_σ, and this implies our assertion because the orders of M_σ and E' are relatively prime. $\square$

13. Prime Action

In this section we consider how, in a maximal subgroup M, a complement of M_σ acts upon M_σ. We establish conditions under which some subgroup X of the complement acts in a *prime* manner on M_σ. Recall that this means that

$$C_{M_\sigma}(g) = C_{M_\sigma}(X) \text{ for all } g \in X^\#$$

or, equivalently,

$$C_{M_\sigma}(P) \subseteq C_{M_\sigma}(X) \text{ for all } P \in \mathcal{E}^1(X),$$

Recall also that X acts *regularly* on M_σ if $C_{M_\sigma}(g) = 1$ for all $g \in X^\#$.

Throughout this section we let M denote an arbitrary maximal subgroup of G. As in Section 12 we also let E be a complement of M_σ in M, and, for $i = 1, 2, 3$, let E_i be Hall $\tau_i(M)$-subgroups of M lying in E such that E_{12} is a Hall $\tau_1(M) \cup \tau_2(M)$-subgroup of E. The basic properties of these subgroups are described at the beginning of Section 12.

Lemma 13.1. Suppose that $M^* \in \mathcal{M}$, $p \in \pi(E) \cap \pi(M^*)$, $p \notin \tau_1(M^*)$, $[M_\sigma \cap M^*, M \cap M^*] \neq 1$, and M^* is not conjugate to M in G. Then

    (a) every p-subgroup of $M \cap M^*$ centralizes $M_\sigma \cap M^*$,
    (b) $p \notin \tau_2(M^*)$, and
    (c) if $p \in \tau_1(M)$, then $p \in \beta(G)$.

Proof. Since $[M_\sigma \cap M^*, M \cap M^*] \subseteq M_\sigma \cap M^{*\prime}$, there exists a prime q in $\sigma(M) \cap \pi(M^{*\prime})$. Let Y be a Sylow q-subgroup of $M^{*\prime}$. Since $M^{*\prime}/M^*_\beta$ is nilpotent by Lemma 10.8, we have $M^*_\beta Y \lhd M^*$, and the Frattini argument yields

$$M^* = N_{M^*}(Y)M^*_\beta.$$

If $p \in \tau_2(M^*)$, then $\mathrm{r}_p(N_{M^*}(Y)) = 2$ and $p \notin \beta(G)$ by Lemma 12.1(g). This contradicts Corollary 12.16(a) and hence proves (b).

Now $p \in \sigma(M^*) \cup \tau_3(M^*)$. Therefore $M^{*\prime}$ contains a Sylow p-subgroup S of M^*. In particular, $p \in \pi(N_{M^*}(Y)')$ if $p \notin \beta(M^*)$. Therefore Corollary 12.16(b) yields (c).

Now let P be any p-subgroup of $M \cap M^*$. Since $M^{*\prime}/M^*_\alpha$ is nilpotent, $M^*_\alpha S \lhd M^*$. Therefore $P \subseteq M^*_\alpha S$. Moreover, $M^*_\alpha S$ is a $\sigma(M)'$-group because $p \in \pi(E)$ and $\sigma(M)$ is disjoint from $\alpha(M^*)$ by Lemma 10.12(a). Thus

$$[M_\sigma \cap M^*, P] \subseteq M^*_\alpha S \cap M_\sigma = 1,$$

and this completes the proof of the lemma. $\square$

Corollary 13.2. Suppose that $p \in \tau_1(M) \cup \tau_3(M)$, P is a nonidentity p-subgroup of M, and $M^* \in \mathscr{M}(N_G(P))$. Then

    (a) every p-subgroup of $M \cap M^*$ centralizes $M_\sigma \cap M^*$,

    (b) every $\tau_1(M^*)'$-subgroup of $E \cap M^*$ centralizes $M_\sigma \cap M^*$, and

    (c) if $[M_\sigma \cap M^*, M \cap M^*] \neq 1$, then $p \in \sigma(M^*)$ and in case $p \in \tau_1(M)$ we even have $p \in \beta(M^*)$.

Proof. Since $p \in \sigma(M^*) \cup \tau_2(M^*)$ by Lemma 12.2(a), M^* is not conjugate to M in G. Now our assertions follow directly from Lemma 13.1. □

Corollary 13.3.

    (a) Every nontrivial cyclic Sylow subgroup of E acts in a prime manner on M_σ.

    (b) The group E_3 acts in a prime manner on M_σ.

Proof. Take $p \in \tau_1(M) \cup \tau_3(M)$ and let P be any nontrivial p-subgroup of E. Take $M^* \in \mathscr{M}(N_G(P))$.

By Corollary 13.2(a), every p-subgroup of $N_E(P)$ centralizes $C_{M_\sigma}(P)$, and this proves (a).

For (b), we can assume that $E_3 \neq 1$ and $p \in \tau_3(M)$ and recall that E_3 lies in E' and is a cyclic normal subgroup of E (Lemma 12.1). Thus $E \subseteq N_G(P) \subseteq M^*$ and $E_3 \subseteq E' \subseteq M^*$ and $E_3 \subseteq E' \subseteq M^{*\prime}$. Therefore $\pi(E_3) \subseteq \pi(M^{*\prime}) \subseteq (\tau_1(M^*))'$. Hence Corollary 13.2(b) shows that E_3 centralizes $C_{M_\alpha}(P)$, and this proves (b). □

The next result is the main step in our investigation of the action of E on M_σ.

Theorem 13.4. Suppose that $p \in \tau_1(M)$, $P \in \mathcal{E}_p{}^1(E)$, $r \in \pi(E)$, and $R \in \mathcal{E}_r{}^1(C_E(P))$. Then $C_{M_\sigma}(P) \subseteq C_{M_\sigma}(R)$.

Proof. Take $q \in \sigma(M)$ and let S be a PR-invariant Sylow q-subgroup of $C_{M_\sigma}(P)$. We have to show that R centralizes S, so assume that

$$Q = [S, R] \neq 1.$$

Take $M^* \in \mathscr{M}(N_G(P))$. Then

$$1 \subset Q = [S, R] \subseteq [M_\sigma \cap M^*, R],$$

and it follows from Corollary 13.2 that

$$p \in \beta(M^*) \text{ and } r \in \tau_1(M^*).$$

Now $1 \subset P \subseteq C_{M_\beta}(RQ)$, and $S = C_S(R) \times Q$ because S is abelian by Theorem 12.13. Thus we can apply Lemma 12.18(a) with r, R, and M^* in place of p, P, and M, to get

$$\mathscr{M}(N_G(Q)) = \{M^*\}.$$

We cannot have case (e) of Proposition 12.15 (with $X = Q$) because $1 \subset P \subseteq M \cap M^*_\sigma$. Therefore $q \in \sigma(M^*)$ by Lemma 10.12(a), and Proposition 12.15(d) shows that $M_\alpha \neq 1$ and

$$r \in \pi(E) \cap \tau_1(M^*) \subseteq \tau_1(M).$$

Since our assumption that $[S, R] \neq 1$ yields $q \notin \alpha(M)$, we can conclude that

$$C_{M_\alpha}(P) \subseteq C_{M_\alpha}(R)$$

and similarly, since $r \in \tau_1(M)$,

$$C_{M_\alpha}(R) \subseteq C_{M_\alpha}(P).$$

It follows that $C_{M_\alpha}(P) = C_{M_\alpha}(R)$. This subgroup is normalized by S because $S \subseteq C_M(P)$, and therefore is centralized by $Q = [S, R]$. Thus $C_{M_\alpha}(R) = C_{M_\alpha}(RQ)$.

Now recall that $\mathcal{M}(N_G(Q)) \neq \{M\}$. Then Lemma 12.18(a) shows that $C_{M_\alpha}(R) \neq C_{M_\alpha}(RQ)$. This contradiction completes the proof of Theorem 13.4. $\square$

Theorem 13.5. Suppose that $E_1 \neq 1$. Then E_1 acts in a prime manner on M_σ.

Proof. Since E_1 is cyclic, this theorem follows from Corollary 13.3 and Theorem 13.4. $\square$

Lemma 13.6. Suppose $1 \subset P \subseteq E_1$, $q \in \sigma(M)$, and $X \in \mathcal{E}_q{}^1(C_{M_\sigma}(P))$, and let S be a Sylow q-subgroup of M_σ. Then

$$\mathcal{M}(C_G(X)) = \mathcal{M}(S) = \{M\}.$$

Proof. By Corollary 12.14, we can assume that

$$q \notin \beta(M) \text{ and } X \not\subseteq M_\sigma'.$$

We can also assume that $P = E_1$ because $C_{M_\sigma}(P) = C_{M_\sigma}(E_1)$ by Theorem 13.5.

Since $q \notin \beta(M)$, by Theorem 12.13, E' centralizes some Sylow q-subgroup of M_σ, and, by Proposition 1.5, we can assume that E normalizes S and that

$$X \subseteq S \subseteq C_{M_\sigma}(E').$$

Moreover, since, by Lemma 12.17, $C_{M_\sigma}(E) \subseteq M_\sigma'$, we know $X \not\subseteq C_{M_\sigma}(E)$ and hence $E_1 E' \neq E$. Therefore $E_2 \neq 1$ because $E_3 \subseteq E'$ and $E = E_1 E_2 E_3$.

Take $p \in \tau_2(M)$ and $Q \in \mathcal{E}_p{}^2(E)$. Then

(13.1) $A \lhd E$ and $C_{M_\sigma}(A) = 1$

by Corollary 12.6(a) and Theorem 12.5(d). Furthermore, Theorem 13.4 shows that $A_0 = C_A(E_1)$ centralizes X. Thus $A = A_0 \times [A, E_1]$ centralizes X, contrary to (13.1). $\square$

Lemma 13.7. Suppose that $E_1 \neq 1$ and that E_1 does not act regularly on E_3. Then $E_1 E_3$ acts in a prime manner on M_σ.

Proof. By assumption, there exist primes p and r such that $P \in \mathcal{E}_p{}^1(E_1)$ centralizes $R \in \mathcal{E}_r{}^1(E_3)$. By Theorem 13.4,

$$C_{M_\sigma}(P) \subseteq C_{M_\sigma}(R).$$

By Theorem 13.5 and Corollary 13.3(b),

(13.2) E_1 and E_3 act in a prime manner on M_σ.

Moreover, E_1 and E_3 have relatively prime orders. Thus, if $C_{M_\sigma}(P) = C_{M_\sigma}(R)$, then $E_1 E_3$ acts in a prime manner on M_σ, as desired.

Henceforth we assume that

(13.3) $$C_{M_\sigma}(P) \subset C_{M_\sigma}(R).$$

We will obtain a contradiction.

Since $1 \subset R \subset E_3$ and $C_{M_\sigma}(R) \neq 1$, Corollary 12.6(d) shows that $\tau_2(M)$ is empty. Thus

(13.4) $$E = E_1 E_2 E_3 = E_1 E_3.$$

As E_3 is a cyclic normal subgroup of E, we know that $R \lhd E$. Take $M^* \in \mathcal{M}(N_G(R))$. Then

$$E \subseteq M^* \text{ and } 1 \subset [C_{M_\sigma}(R), P] \subseteq [M_\sigma \cap M^*, E_1].$$

By (13.2), $C_{E_1}(M_\sigma \cap M^*) = 1$. Thus, by Corollary 13.2(b), $\pi(E_1) \subseteq \tau_1(M^*)$.

Now E_1 is contained in some Hall $\tau_1(M^*)$-subgroup E_1^* of M^*. By Corollary 13.2(c), $R \subseteq M_\sigma^*$. Therefore

$$1 \subset P \subseteq C_{E_1^*}(R).$$

Since E_1^* is prime on M_σ^* by Theorem 13.5, E_1^* centralizes R. In particular, E_1 centralizes R. By (13.4), $R \subseteq Z(E)$, but, by Lemma 12.1, $C_{E_3}(E) = 1$, a contradiction. This completes the proof of Lemma 13.7. $\square$

Lemma 13.8. The following configuration is impossible:

(1) $M^* \in \mathcal{M}$ and M^* is not conjugate to M in G,
(2) $p \in \tau_1(M) \cap \tau_1(M^*)$ and $P \in \mathcal{E}_p{}^1(M \cap M^*)$,
(3) Q and Q^* are P-invariant Sylow subgroups (possibly for different primes) of $M \cap M^*$,
(4) $C_Q(P) = 1$ and $C_{Q^*}(P) = 1$,
(5) $N_G(Q) \subseteq M^*$ and $N_G(Q^*) \subseteq M$.

Proof. Assume this configuration and note the symmetry between M and M^*.

By (3), (5), and the Uniqueness Theorem, Q is a nonidentity Sylow subgroup of M for a prime $q \notin \alpha(M)$. Since $P \subseteq M \cap M^*$ and $C_Q(P) = 1$,

$$Q = [Q, P] \subseteq M' \cap M^{*\prime}.$$

Thus $QM_\alpha \lhd M$, and $M = N_M(Q)M_\alpha$ because M'/M_α is nilpotent by Theorem 10.2.

By Lemma 12.18, $C_{M_\beta}(P) \neq 1$ and $C_{M^*_\beta}(P) \neq 1$. Furthermore, by Proposition 10.14(d), $N_G(X) \subseteq M$ for every nonidentity $\beta(M)$-subgroup X of $C_M(P)$, and similarly for M^* and every conjugate M^g of M.

Let H be a Hall $(\beta(M) \cup \beta(M^*))$-subgroup of $C_G(P)$ and take any $s \in \pi(F(H))$ and $t \in \pi(F(C_{M_\beta}(P)))$. By the symmetry between M and M^*, we can assume that $s \in \beta(M)$ and then that $H \supseteq C_{M_\beta}(P)$. Let $X = \mathcal{O}_s(H)$ and $Y = \mathcal{O}_t(C_{M_\beta}(P))$. Then $X \subseteq M^g$ for some $g \in G$. It follows that $M^g \supseteq N_G(X) \supseteq H \supseteq Y$ and $M \supseteq N_G(Y) \supseteq C_G(Y)$. Since $Y \subseteq M \cap M^g$, Theorem 10.1(b) yields

$$M^g = M^h \text{ for some } H \in C_G(Y) \subseteq M.$$

Thus $M = M^g \supseteq H$.

Take $r \in \beta(M^*) \cap \pi(H)$. Then r divides $|C_M(P)|$. By Lemma 10.12(a), $r \notin \sigma(M)$.

Since $M = N_M(Q)M_\alpha$ and $r \in \pi(C_M(P))$, some subgroup $R \subseteq N_M(Q)$ of order r is centralized by P. Then $R \subseteq N_G(Q) \subseteq M^*$ and consequently $N_G(R) \subseteq M^*$ by Proposition 10.14(d). Now, since PR is conjugate in M to an abelian subgroup of E, Theorem 13.4 yields

$$1 \subset X \subseteq C_{M_\sigma}(P) \subseteq C_{M_\sigma}(R) \subseteq M^*.$$

Then $[X, Q] \subseteq [M_\alpha \cap M^*, Q] \subseteq M^*_\alpha$ because $Q \subseteq M^{*\prime}$, $M^{*\prime}/M^*_\alpha$ is nilpotent and $M_\alpha \cap M^*$ is a Q-invariant q'-subgroup of M^*. On the other hand, $[X, Q] \subseteq M_\alpha$ and $M_\alpha \cap M^*_\alpha = 1$ by Lemma 10.12. Thus $[X, Q] = 1$ and $X \subseteq C_{M_\alpha}(PQ)$, contrary to the fact that $C_{M_\alpha}(PQ) = 1$ by Lemma 12.18. $\square$

Theorem 13.9. Suppose $M^* \in \mathcal{M}$ and M^* is not conjugate to M in G. Then $\sigma(M)$ is disjoint from $\sigma(M^*)$.

Proof. Suppose $q \in \sigma(M) \cap \sigma(M^*)$. Let S be an E-invariant Sylow q-subgroup of M_σ. Replacing M^* by some conjugate if necessary, we can assume that S is also a Sylow q-subgroup of M^*. Then

$$N_G(S) \subseteq M \cap M^*.$$

By Corollary 12.6(f), $\tau_2(M)$ is empty. Therefore, by Lemma 12.1(c), $\tau_1(M)$ is not empty.

Take $p \in \tau_1(M)$ and $P \in \mathcal{E}_p{}^1(E_1)$. By Lemma 13.6, $C_S(P) = 1$. Therefore, by Lemma 13.1(a), $p \in \tau_1(M^*)$. Now Lemma 13.8, applied with $Q = Q^* = S$, yields a contradiction. $\square$

Theorem 13.10. Suppose some $P \in \mathcal{E}_p{}^1(E_1)$ does not centralize E_3. Then

    (a) E_1 acts regularly on E_3,
    (b) E_3 acts regularly on M_σ, and
    (c) $C_{M_\sigma}(P) \neq 1$.

Proof. Since E_3 is a cyclic normal Hall subgroup of E, there exists $q \in \tau_3(M)$ such that P acts regularly on the Sylow q-subgroup Q of E_3. Thus

$$(13.5) \qquad Q = C_Q(P)[Q, P] = [Q, P] \subseteq E'.$$

Take $M^* \in \mathcal{M}(N_G(Q))$. By Lemma 12.2(b), M^* is not conjugate to M in G. In particular, $M^* \neq M$. Consequently, by Lemma 12.18,

$$C_{M_\alpha}(P) \neq 1 \text{ and } C_{M_\alpha}(PQ) = 1.$$

Since $M_\alpha \subseteq M_\sigma$, this proves (c) and shows that $E_1 E_3$ is not prime on M_σ. Therefore (a) follows from Lemma 13.7.

Now assume that (b) is false. Then $C_{M_\sigma}(E_3) \neq 1$ because E_3 is prime on M_σ by Corollary 13.3(b). Take $q^* \in \pi(C_{M_\sigma}(E_3))$ and let Q^* be an E-invariant Sylow q^*-subgroup of $C_{M_\sigma}(E_3) = C_{M_\sigma}(Q)$. Since Q centralizes $M_\sigma \cap M^*$ by Corollary 13.2(a), Q^* is a Sylow q^*-subgroup of $M_\sigma \cap M^*$ and hence of $M \cap M^*$.

By (13.5) and Lemma 12.19, $q^* \in \beta(M)$ or Q^* is a Sylow q^*-subgroup of M_σ. Hence, by Proposition 10.14(d) and the definition of $\sigma(M)$,

$$N_G(Q^*) \subseteq M.$$

If $q^* \in \beta(M)$, then

$$C_{Q^*}(P) \subseteq C_{M_\alpha}(PQ) = 1,$$

and, on the other hand, if $q^* \notin \beta(M)$, then $C_{Q^*}(P) = 1$ by Lemma 13.6 because $\mathcal{M}(Q^*) \neq \{M\}$. In particular, $[M_\sigma \cap M^*, P] \neq 1$. Therefore, by Lemma 13.1(a), $p \in \tau_1(M^*)$. Now Lemma 13.8 yields a contradiction. $\square$

Corollary 13.11. Suppose $E_3 \neq 1$ and E_3 does not act regularly on M_σ. Then

    (a) $E_1 \neq 1$,
    (b) $E = E_1 E_3$,
    (c) E acts in a prime manner on M_σ, and
    (d) every $X \in \mathcal{E}^1(E)$ is normal in E.

Proof. By Corollary 12.6(d), $\tau_2(M)$ is empty. By Lemma 12.1, this implies (b) and (a). Moreover Theorem 13.10(b) shows that every $P \in \mathcal{E}_p{}^1(E_1)$ centralizes E_3. By (b), this implies (d) because $E_3 \triangleleft E$ and both E_1 and E_3 are cyclic. Furthermore, E_1 does not act regularly on E_3. Hence Lemma 13.7 yields (c). $\square$

Finally we turn to the primes in $\tau_2(M)$.

Lemma 13.12. Suppose $p \in \tau_1(M)$, $P \in \mathcal{E}_p{}^1(E)$, $q \in \tau_2(M)$, $A \in \mathcal{E}_q{}^2(E)$, and $C_A(P) \neq 1$. Then $C_{M_\sigma}(P) = 1$.

Proof. Suppose that $C_{M_\sigma}(P) \neq 1$. Then $A \lhd E$ and $P \not\subseteq C_E(A)$ by Corollary 12.6(a) and (e). Therefore $Y = C_A(P)$ has order q. By Theorem 13.4,

$$1 \subset C_{M_\sigma}(P) \subseteq C_{M_\sigma}(Y).$$

Therefore, by Corollary 12.6(c),

$$\mathcal{M}(C_G(Y)) = \{M\}.$$

For $M^* \in \mathcal{M}(N_G(A))$ we have $q \in \sigma(M^*)$ and $p \in \tau_1(M^*) \cup \tau_2(M^*)$ by Lemma 12.11.

Suppose that $p \in \tau_2(M^*)$. Then, by Corollary 12.6(c) applied to M^*, $\mathcal{M}(C_G(P)) = \{M^*\}$ because $1 \subset C_A(P) \subseteq C_{M_\sigma^*}(P)$. Hence

$$1 \subset C_G(P) \cap M_\sigma \subseteq M^* \cap M_\sigma,$$

contrary to Theorem 12.5(e). Thus $p \in \tau_1(M^*)$.

Since $Y \in \mathcal{E}_q{}^1(C_{M_\sigma^*}(P))$, it follows from Lemma 13.6 applied to M^* that $\mathcal{M}(C_G(Y)) = \{M^*\}$, a contradiction. $\square$

Lemma 13.13. Suppose $p \in \tau_1(M) \cup \tau_3(M)$, $P \in \mathcal{E}_p{}^1(E)$, and $C_{M_\sigma}(P) \neq 1$. Then for every $M^* \in \mathcal{M}(N_G(P))$ we have $p \in \sigma(M^*)$.

Proof. By Lemma 12.2, $P \in \sigma(M^*) \cup \tau_2(M^*)$. Suppose $p \in \tau_2(M^*)$. We will obtain a contradiction.

Choose $q \in \pi(C_{M_\sigma}(P))$ and $Q \in \mathcal{E}_q{}^1(C_{M_\sigma}(P))$. By Theorem 13.9, we know that $q \notin \sigma(M^*)$. Let E^* be a complement of M_σ^* in M^* that contains PQ. Take $A \in \mathcal{E}_p{}^2(E^*)$. By Corollary 12.6(a), $A \lhd E^*$ and $P \subseteq A$.

We can assume that P lies in E_1 or in E_3. If $P \subseteq E_3$, then parts (a) and (c) of Corollary 13.11 show that $E_1 \neq 1$ and $C_{M_\sigma}(P) = C_{M_\sigma}(E_1)$ because $1 \subset Q \subseteq C_{M_\sigma}(P)$. Therefore, in any case, we have $C_G(Q) \subseteq M$ by Lemma 13.6, and this implies that $A \not\subseteq C_{E^*}(Q)$ because $r_p(M) = 1$. Therefore, by Corollary 12.10(c), $q \in \tau_1(M^*)$. Furthermore,

$$1 \subset P = C_A(Q).$$

Consequently, by Lemma 13.12 applied to M^*,

$$C_{M_\sigma^*}(Q) = 1.$$

However $P = C_A(Q)$, and Corollary 12.9(c) shows that if $C_{M_\sigma^*}(Q) = 1$, then $N_G(P) \not\subseteq M^*$, a contradiction. $\square$

CHAPTER IV

The Family of All Maximal Subgroups of G

In the previous chapter we considered mainly a single maximal subgroup of G in isolation. In this chapter we exploit also the interrelationships among the maximal subgroups of G. Our first main application of these interrelationships shows that a particular family of maximal subgroups of G is either empty or consists of precisely two conjugacy classes of maximal subgroups of G (Theorem 14.7). Eventually, we obtain our main results in Section 16.

14. Maximal Subgroups of Type $\mathscr{P}$ and Counting Arguments

As a consequence of the preceding section, $\sigma(M) \cap \sigma(H)$ is empty and $M_\sigma \cap H_\sigma$ is trivial for nonconjugate subgroups $M, H \in \mathscr{M}$ (Theorem 13.9). Clearly every $p \in \pi(G)$ appears in some set $\sigma(M)$ because some $M \in \mathscr{M}$ contains the normalizer of a Sylow p-subgroup of G. Therefore the sets $\sigma(M)$, where M ranges over a set of representatives of the conjugacy classes of $\mathscr{M}$ under G, form a partition of $\pi(G)$.

Let $\sigma_1, \ldots, \sigma_s$ denote the sets $\sigma(M)$ for $M \in \mathscr{M}$. Then every element $g \in G$ can be written uniquely as a product

$$g = g_1 \cdots g_s$$

of pairwise commuting σ_i-elements g_i. Disregarding the ordering and factors $g_i = 1$, we call this the σ-*decomposition* of g, and the number $\ell_\sigma(g)$ of nonidentity factors g_i may be called the σ-*length* of g. Clearly each subgroup containing g contains all the g_i because $g_i \in \langle g \rangle$ for each i.

For every subset X of G, define

$$\mathscr{M}_\sigma(X) = \{ M \in \mathscr{M} \mid X \subseteq M_\sigma \}.$$

For $g \in G$, let $\mathscr{M}_\sigma(g)$ be $\mathscr{M}_\sigma(\{g\})$. By Corollary 12.16, every $\sigma(M)$-element, where $M \in \mathscr{M}$, is conjugate to an element of M_σ. Thus $\ell_\sigma(g) \leq 1$

means that $\mathcal{M}_\sigma(g)$ is not empty. In Theorem 14.4 we will show that $C_G(g)$ acts transitively on $\mathcal{M}_\sigma(g)$ if $\ell_\sigma(g) = 1$.

Now we define a further subset of $\pi(M)$ for $M \in \mathcal{M}$ and several families of maximal subgroups of G:

$$\kappa(M) = \left\{ p \in \tau_1(M) \cup \tau_3(M) \mid C_{M_\sigma}(P) \neq 1 \text{ for some } P \in \mathcal{E}_p{}^1(M) \right\},$$

$$\mathcal{M}_{\mathscr{F}} = \{ M \in \mathcal{M} \mid \kappa(M) \text{ is empty} \},$$

$$\mathcal{M}_{\mathscr{P}} = \{ M \in \mathcal{M} \mid \kappa(M) \text{ is not empty} \},$$

$$\mathcal{M}_{\mathscr{P}_1} = \{ M \in \mathcal{M}_{\mathscr{P}} \mid \kappa(M) = \pi(M) - \sigma(M) \},$$

$$\mathcal{M}_{\mathscr{P}_2} = \{ M \in \mathcal{M}_{\mathscr{P}} \mid \kappa(M) \neq \pi(M) - \sigma(M) \}.$$

Note that for $p \in \kappa(M)$ we know $\mathrm{r}_p(M) = 1$, and every $P \in \mathcal{E}_p{}^1(M)$ satisfies $C_{M_\sigma}(P) \neq 1$ because all these P are conjugate in M.

Though not explicitly, the family $\mathcal{M}_{\mathscr{P}}$ has already been the subject of Section 13 ($\mathscr{P}$ stands for *proper prime action*). It will be investigated more closely in this section. Maximal subgroups in the complementary family $\mathcal{M}_{\mathscr{F}}$ will turn out to be of *Frobenius type* in the sense of Section 16. Actually, this will follow immediately from Theorem 12.12 and the last assertion of the following lemma.

Lemma 14.1. Suppose that $M \in \mathcal{M} - \mathcal{M}_{\mathscr{P}_1}$. Take any prime $p \in \pi(M) - (\sigma(M) \cup \kappa(M))$, let S be a Sylow p-subgroup of M, and let $A = \Omega_1(S)$. Then $|A| \leq p^2$, $C_{M_\sigma}(A) = 1$, and M_σ is nilpotent.

Proof. If $p \in \tau_2(M)$, then the assertions of the lemma follow directly from Theorem 12.5(a), (b), and (d).

If $p \in \tau_1(M) \cap \tau_3(M)$, then $|A| = p$ and $C_{M_\sigma}(A) = 1$ because $p \notin \kappa(M)$. Now Theorem 3.7 shows that M_σ is nilpotent. $\square$

Recall that a group in which all of the Sylow subgroups are cyclic is called a *Z-group*, and that a nonempty subset X of G is a *TI-subset* of G if $X \cap X^g \subseteq 1$ for all $x \in G - N(X)$. In particular, a nonidentity proper subgroup H of G is a *TI-subgroup* of G if $H \cap H^g = 1$ for all $g \in G - N(H)$.

The following proposition contains nearly everything that we have proved in Section 13.

Proposition 14.2. Suppose $M \in \mathcal{M}_{\mathscr{P}}$. Let K be a Hall $\kappa(M)$-subgroup of M and define $K^* = C_{M_\sigma}(K)$. Then the following conditions are satisfied.

  (a) The group K acts in a prime manner on M_σ and acts regularly on some abelian Hall $(\kappa(M) \cup \sigma(M))'$-subgroup U of M. (Thus $U M_\sigma$ is a normal complement of K in M).
  (b) For every $X \in \mathcal{E}^1(K)$,
       (1) $N_M(X) = N_M(K) = K \times K^*$, and
       (2) $X \subseteq M_\sigma^*$ for each $M^* \in \mathcal{M}(N_G(X))$.
  (c) $K^* \neq 1$ and every $X \in \mathcal{E}^1(K^*)$ satisfies $\mathcal{M}(C_G(X)) = \{M\}$.

(d) Every $g \in G - M$ satisfies $K^* \cap M^g = 1$ and every $g \in M - (K \times K^*)$ satisfies $K \cap K^g = 1$.

(e) For every prime $p \in \pi(K^*)$ and every Sylow p-subgroup S of M_σ, $\mathscr{M}(S) = \{M\}$ and $S \not\subseteq K^*$.

(f) Every $\sigma(M)$-subgroup Y of G satisfying $Y \cap K^* \neq 1$ lies in M_σ.

(g) If $M \in \mathscr{M}_{\mathscr{P}_2}$, i.e., $U \neq 1$, then $\sigma(M) = \beta(M)$, K has prime order, and M_σ is a nilpotent TI-subgroup of G.

Proof. Take E, E_1, E_2, and E_3, as in Section 12, such that $E \supseteq K$. Thus E is a complement of M_σ in M and E_i is a Hall $\tau_i(M)$-subgroup of E (and of M) for $i = 1, 2, 3$. By Lemma 12.1,

$$E_2 E_3 \lhd E = E_1 E_2 E_3 \text{ and } E_1 \text{ is cyclic.}$$

We first prove parts (a) and (b1).

If $\kappa(M) \cap \tau_3(M)$ is not empty, which means that $E_3 \neq 1$ and E_3 does not act regularly on M_σ, then Corollary 13.11 yields $E_1 \neq 1$, $E = E_1 E_3$, E is prime on M_σ, and every $X \in \mathcal{E}^1(E)$ is normal in E. Then $K = E$, and (a) and (b1) are clear (let $U = 1$).

If $\kappa(M) \subseteq \tau_1(M)$, then $\kappa(M) = \tau_1(M)$ because E_1 is prime on M_σ by Theorem 13.5, and thus we may (and shall) assume that $K = E_1$. Then $K = E_1$ acts regularly on $U = E_2 E_3$ by Lemma 13.12 and Lemma 13.7. Now $U = [U, K] = E'$, and E' is abelian by Corollary 12.10(b). Again (a) and (b1) hold.

Now, in any case, since K acts in a prime manner, but not regularly, on M_σ, K^* is not trivial. Hence (b2) and (c) follow from Lemma 13.13 and Lemma 13.6.

Suppose $g \in G$ and $X \in \mathcal{E}^1(K^* \cap M^g)$. By (c), $C_G(X) \subseteq M$. Therefore, by Theorem 10.1(a), $g \in M$. This proves the first assertion of (d), and the second assertion follows easily from (b1) since K is a Z-group.

For $X \in \mathcal{E}^1(K)$, $N_G(X) \not\subseteq M$ by (b2). In particular, $\mathscr{M}(K^*) \neq \{M\}$, and then (e) follows directly from Lemma 13.6.

For Y as in (f), Corollary 12.16 yields an element $g \in G$ such that $Y \subseteq M^g$. By (d), $g \in M$. Therefore $Y \subseteq M_\sigma$.

Finally assume $U \neq 1$ with U as above. Then E is a Frobenius group with Frobenius kernel U. By Lemma 14.1, $C_{M_\sigma}(U) = 1$ and M_σ is nilpotent. Since K is prime on M_σ, it follows from Theorem 3.10(a) that K has prime order. Clearly $U = [U, K] = E'$. Therefore, by Lemma 12.19, U centralizes a Hall $\beta(M)'$-subgroup of M_σ. Consequently, since $\beta(M) \subseteq \sigma(M)$ and $C_{M_\sigma}(U) = 1$, we have $\beta(M) = \sigma(M)$. By Lemma 12.17, $M_\sigma \cap M_\sigma{}^g$ is a $\beta(M)'$-group for every $g \in G - M$. Thus $M_\sigma \cap M_\sigma{}^g = 1$ for every $g \in G - M$. This proves (g). $\square$

Corollary 14.3. Suppose $M \in \mathcal{M}$, $x \in M_\sigma^\#$, and x' is a nonidentity $\sigma(M)'$-element of $C_M(x)$. Then

(1) $\pi(\langle x' \rangle) \subseteq \kappa(M)$ and $C_G(x) \subseteq M$, or
(2) $\pi(\langle x' \rangle) \subseteq \tau_2(M)$, $\ell_\sigma(x') = 1$, and $\mathcal{M}(C_G(x')) = \{M\}$.

Proof. Suppose that $\pi(\langle x' \rangle)$ contains a prime number $p \in \tau_2(M)'$. Then $p \in \tau_1(M) \cup \tau_3(M)$ and $X \in \mathcal{E}_p{}^1(\langle x' \rangle)$ satisfies

$$C_{M_\sigma}(X) \supseteq C_{M_\sigma}(x') \supseteq \langle x \rangle \supset 1.$$

Thus $p \in \kappa(M)$ and X lies in some Hall $\kappa(M)$-subgroup K of M. Therefore, by Lemma 14.1(b),

$$C_M(x') \subseteq C_M(X) \subseteq K \times C_{M_\sigma}(K).$$

Consequently $x' \in K$, $x \in C_{M_\sigma}(K)$, and then Proposition 14.2(c) yields $C_G(x) \subseteq M$.

Assume next that x' is a $\tau_2(M)$-element. Since $C_{M_\sigma}(x') \neq 1$, Corollary 12.10(e) shows that $\mathcal{M}(C_G(x')) = \{M\}$ and, by Lemma 12.11(a), $\ell_\sigma(x') = 1$ because $\tau_2(M) \subseteq \sigma(M^*)$ for some $M^* \in \mathcal{M}$. $\square$

Theorem 14.4. Suppose $x \in G^\#$ and $\mathcal{M}_\sigma(x)$ is not empty. Then $C_G(x)$ has a normal Hall subgroup $R(x)$ that acts sharply transitively on $\mathcal{M}_\sigma(x)$.

Furthermore, if $|\mathcal{M}_\sigma(x)| > 1$, then $C_G(x)$ lies in a unique subgroup $N = N(x) \in \mathcal{M}$ and, for every $M \in \mathcal{M}_\sigma(x)$,

(a) $R(x) = C_{N_\sigma}(x) \supset 1$,
(b) $C_G(x) = C_{M \cap N}(x) R(x)$,
(c) $\pi(\langle x \rangle) \subseteq \tau_2(N) \subseteq \sigma(M)$,
(d) $\pi(M) \cap \sigma(N) \subseteq \beta(N)$,
(e) $M \cap N$ is a complement of N_σ in N, and
(f) **(D. Sibley)** $N \in \mathcal{M}_F \cup \mathcal{M}_{\mathcal{P}_2}$.

Proof. If $|\mathcal{M}_\sigma(x)| = 1$, we can let $R(x) = 1$. So henceforth we assume that $|\mathcal{M}_\sigma(x)| \geq 2$.

Take $M \in \mathcal{M}_\sigma(x)$, $q \in \pi(\langle x \rangle)$, $X \in \mathcal{E}_q{}^1(\langle x \rangle)$, and $N \in \mathcal{M}(N_G(X))$. Let $R(x) = C_{N_\sigma}(x)$, as in (a). Then

$$C_G(x) \subseteq N_G(\langle x \rangle) \subseteq N_G(X) \subseteq N$$

and

$$\mathcal{M}_\sigma(x) \subseteq \mathcal{M}_\sigma(X).$$

Thus $R(x)$ is a normal Hall subgroup of $C_G(x)$.

By Theorem 13.9 and Theorem 10.1(b), $C_G(X)$ acts transitively on $\mathcal{M}_\sigma(X)$. In particular, $C_G(X) \not\subseteq M$ and $N \neq M$.

Thus we can apply Proposition 12.15 because $\sigma(M)$ is disjoint from $\sigma(M^*)$ by Theorem 13.9. It follows that $q \in \tau_2(N)$ and that (d) and (e) of the present proposition hold. Since $\tau_2(N)$ is not empty and, by the definition of $\kappa(N)$, $\kappa(N) \subseteq \tau_1(N) \cap \tau_3(N)$, we know $N \notin \mathscr{M}_{\mathscr{P}_1}$. Thus (f) holds.

If $L \in \mathscr{M}_\sigma(x)$, then, by the previous paragraph, $L = M^u$ for some $u \in C_G(X) = C_N(X)$, and we can choose u in N_σ because $N = (M \cap N)N_\sigma$. Then

$$M^{u^x} = M^{x^{-1}ux} = M^{ux} = L^x = L = M^u.$$

However, if $M^{u'} = M^u$ for any element $u' \in N_{\sigma'}$, then it follows that $u'u^{-1} \in N_G(M) \cap N_\sigma = M \cap N_\sigma = 1$. Thus $u^x = u$ and $u \in C_{N_\sigma}(x) = R(x)$. Moreover, $R(x) = C_{N_\sigma}(x)$ acts sharply transitively on $\mathscr{M}_\sigma(x)$. Therefore $C_G(x) = (C_G(x) \cap N_G(M))R(x) = C_M(x)R(x)$ and $R(x) = C_{N_\sigma}(x) \supset 1$. This proves (b) because $C_G(x) \subseteq N$.

Since $\kappa(N) \subseteq \tau_1(N) \cup \tau_3(N)$ and $q \in \pi(\langle x \rangle) \cap \tau_2(N)$, x is not a $\kappa(N)$-element. By Corollary 14.3, $\pi(\langle x \rangle) \subseteq \tau_2(N)$, and $\mathscr{M}(C_G(x)) = \{N\}$ because $C_{N_\sigma}(x) \neq 1$.

For each $p \in \tau_2(N)$, (e) and Corollary 12.6(a) yield some $A \in \mathcal{E}_p^{\,2}(N)$ that is normal in $M \cap N$. It follows that $N_{M_\sigma}(A) \supseteq \langle x \rangle \supset 1$. Since $r_p(M) \geq 2$, $p \in \sigma(M) \cap \tau_2(M)$. If $p \in \tau_2(M)$, then $N_{M_\sigma}(A) = C_{M_\sigma}(A) = 1$ by Corollary 12.6(b). Thus $p \in \sigma(M)$, and this yields (c) and completes the proof of the theorem. $\square$

In the following we use the groups $R(x)$ defined in Theorem 14.4 for every $x \in G$ of σ-length 1.

Furthermore, for each $M \in \mathscr{M}$ we let

$$\widetilde{M} = \{\, xx' \mid x \in M_\sigma{}^\# \text{ and } x' \in R(x) \,\}.$$

Note that $g = xx'$ is the σ-decomposition of such a product $g \in \widetilde{M}$. In particular, $\ell_\sigma(g) \leq 2$ for each $g \in \widetilde{M}$.

Recall, as in Section 1, for any subset T of G we define

$$\mathscr{C}_G(T) = \{\, t^g \mid t \in T \text{ and } g \in G \,\}.$$

Lemma 14.5.

(a) If x and y are distinct elements of G of σ-length one, then $xR(x)$ is disjoint from $yR(y)$.

(b) If M_1, $M_2 \in \mathscr{M}$ and M_2 is not conjugate to M_1 in G, then $\widetilde{M}_2$ is disjoint from $\widetilde{M}_1$.

(c) If $M \in \mathscr{M}$, then $|\mathscr{C}_G(\widetilde{M})| = (|M_\sigma| - 1)|G : M|$.

Proof. (a) Suppose $g = xx' \in yR(y)$ with $x' \in R(x)$ and $x \neq y$. Then $x' = y$ and $x \in R(y)$ because y is a factor of the σ-decomposition of g. Take $N \in \mathscr{M}(C_G(x))$ and $M \in \mathscr{M}(C_G(y))$. By Theorem 14.4, $x \in R(y) \subseteq M_\sigma$, $y \in R(x) \subseteq N_\sigma$, and $y \in N_\sigma \cap M = 1$, a contradiction.

(b) By Theorem 13.9, $M_{1\sigma} \cap M_{2\sigma} = 1$. Thus, by (a), $\widetilde{M}_1 \cap \widetilde{M}_2$ is empty.

(c) With x ranging over $\mathscr{C}_G(M_\sigma{}^{\#})$, we have

$$\mathscr{C}_G(\widetilde{M}) = \bigcup_{x \in \mathscr{C}_G(M_\sigma{}^{\#})} xR(x)$$

and therefore, by (a),

$$|\mathscr{C}_G(\widetilde{M})| = \sum_{x \in \mathscr{C}_G(M_\sigma{}^{\#})} |R(x)|.$$

Next we count the number n of all pairs (x, M^g) with $g \in G$ and $x \in M_\sigma{}^g$. By Theorem 14.4, each $x \in \mathscr{C}_G(M_\sigma{}^{\#})$ belongs to exactly $|R(x)|$ subgroups M^g. Thus

$$n = \sum_{x \in \mathscr{C}_G(M_\sigma{}^{\#})} |R(x)|.$$

On the other hand

$$n = |M_\sigma{}^{\#}||\{M^g \mid g \in G\}| = |M_\sigma{}^{\#}||G : N_G(M)| = |M_\sigma{}^{\#}||G : M|. \quad \square$$

Lemma 14.6. Each element $g \in G^{\#}$ satisfies exactly one of the following two conditions:

  (1) $g = xx'$ with $\ell_\sigma(x) = 1$ and $x' \in R(x)$,
  (2) $g = yy'$ with $\ell_\sigma(y) = 1$ and y' a nonidentity $\kappa(M)$-element of $C_M(y)$ for some $M \in \mathscr{M}_\sigma(y)$.

Proof. Suppose (1) and (2) hold. Since y' is a $\sigma(M)'$-element of $C_G(y)$, y is a factor of the σ-decomposition of g. Thus $y = x$ or $y = x'$. In particular, $x' \neq 1$. By Corollary 14.3, $M \in \mathscr{M}(C_G(y))$, and, by Theorem 14.4(c), x is a $\tau_2(N)$-element for the unique $N \in \mathscr{M}(C_G(x))$. Thus $y \neq x$. Therefore $y = x'$ and $y' = x$. Now $x' \in R(x) = C_{N_\sigma}(x) \subseteq N_\sigma$ and $y = x' \in M_\sigma \cap N_\sigma$. Then M and N are conjugate in G by Theorem 13.9, contrary to the fact that $y' = x$ is a $\kappa(M)$-element and a $\tau_2(N)$-element.

Assume next that g satisfies neither of our two conditions. It follows that $\ell_\sigma(g) > 1$ because (1) is false. Let x be a factor (of σ-length 1) of the σ-decomposition of g. Take $M \in \mathscr{M}_\sigma(x)$ and $N \in \mathscr{M}(C_G(x))$, and write $g = xx'$.

If $g \in M$, then x' is a $\sigma(M)'$-element of M, but not a $\kappa(M)$-element because (2) is false. Then Corollary 14.3 yields $\ell_\sigma(x') = 1$ and $\mathscr{M}(C_G(x')) = \{M\}$. Therefore $x \in C_{M_\sigma}(x') = R(x')$, contrary to our assumption that (1) is false.

This proves that $g \notin M$, hence $C_G(x) \not\subseteq M$ and, by the same argument, no factor of the σ-decomposition of g lies in N_σ. Therefore g is a $\sigma(N)'$-element of N. Consequently, because $M \cap N$ is a complement of N_σ in N by Theorem 14.4(e), $g \in (M \cap N)^a$ for some $a \in N$. Then $g \in M^a$ and

$x \in M_\sigma{}^a$, and we might have chosen M^a instead of M, contrary to the fact that $g \notin M$. This completes the proof of the lemma. $\square$

In the remainder of this section we will extend our results on individual members of $\mathscr{M}_\mathscr{P}$, but we also gain very explicit information on the family $\mathscr{M}_\mathscr{P}$ as a whole and the role it plays for the global structure of G.

Theorem 14.7. Suppose $M \in \mathscr{M}_\mathscr{P}$ and K is a Hall $\kappa(M)$-subgroup of M. Let $K^* = C_{M_\sigma}(K)$, $k = |K|$, $k^* = |K^*|$, $Z = K \times K^*$, and $\widehat{Z} = Z - (K \cup K^*)$. Then, for some other $M^* \in \mathscr{M}_\mathscr{P}$ not conjugate to M,

(a) $\mathscr{M}(C_G(X)) = \{M^*\}$ for every $X \in \mathcal{E}^1(K)$,
(b) K^* is a Hall $\kappa(M^*)$-subgroup of M^* and a Hall $\sigma(M)$-subgroup of M^*,
(c) $K = C_{M_\sigma^*}(K^*)$ and $\kappa(M) = \tau_1(M)$,
(d) Z is cyclic and for every $x \in K^\#$ and $y \in K^{*\#}$, $M \cap M^* = Z = C_M(x) = C_{M^*}(y) = C_G(xy)$,
(e) $\widehat{Z}$ is a TI-subset of G with $N_G(\widehat{Z}) = Z$, $\widehat{Z} \cap M^g$ empty for all $g \in G - M$, and

$$|\mathscr{C}_G(\widehat{Z})| = \left(1 - \frac{1}{k} - \frac{1}{k^*} + \frac{1}{kk^*}\right)|G| > \frac{1}{2}|G|,$$

(f) M or M^* lies in $\mathscr{M}_{\mathscr{P}_2}$ and, accordingly, K or K^* has prime order,
(g) every $H \in \mathscr{M}_\mathscr{P}$ is conjugate to M or M^* in G, and
(h) M' is a complement of K in M.

Proof. Let $M_1, \ldots, M_n$ be the distinct maximal subgroups of G containing $N_G(X)$ for some $X \in \mathcal{E}^1(K)$, say,

$$M_i \in \mathscr{M}(N_G(X_i)) \text{ for each } X_i \in \mathcal{E}^1(K).$$

Now we will examine M_i for some arbitrary choice of i. By Proposition 14.2(b),

$$Z = K \times K^* \subseteq M_i \text{ and } X_i \subseteq M_{i\sigma}.$$

In particular, M_i is not conjugate to M in G because $\pi(X_i) \subseteq \kappa(M) \subseteq \sigma(M)'$. Therefore, by Theorem 13.9, $\sigma(M)$ is disjoint from $\sigma(M_i)$. In particular, K^* is a $\sigma(M_i)'$-subgroup of M_i.

Take $X^* \in \mathcal{E}^1(K^*)$. By Proposition 14.2(c),

$$\mathscr{M}(C_G(X^*)) = \{M\} \neq \{M_i\}.$$

Thus, by Corollary 14.3, $\pi(X^*) \subseteq \kappa(M_i)$ because $C_{M_{i\sigma}}(X^*) \supseteq X_i \supset 1$. Therefore, because X^* is an arbitrary element of $\mathcal{E}^1(K^*)$,

$$\pi(K^*) \subseteq \kappa(M_i).$$

Moreover, $M \supseteq N_G(X^*)$.

Let K_i be a Hall $\kappa(M_i)$-subgroup of M_i that contains X^* and define $K_i^* = C_{M_{i\sigma}}(K_i)$. By Proposition 14.2(b),

$$N_{M_i}(X^*) = K_i \times K_i^*,$$

and it follows that $X_i \subseteq K_i^*$ and $K \subseteq K_i \times K_i^*$.

Since we could have chosen K_i subject to $K^* \subseteq K_i$, we know K^* lies in $N_{M_i}(X^*)$. Thus $K \times K^* \subseteq K_i \times K_i^*$. Similarly, with $(M_i, K_i, X^*, M, K, X_i)$ in place of $(M, K, X_i, M_i, K_i, X^*)$, we have $K_i \times K_i^* \subseteq K \times K^*$ because $M \supseteq N_G(X^*)$ above. Thus

$$Z = K \times K^* = K_i \times K_i^*.$$

Now let $M_0 = M$, $K_0 = K$, and $K_0^* = K^*$. By Proposition 14.2(c), applied to each M_i $(i = 0, 1, \ldots, n)$,

$$K_i^* \cap K_j^* = 1 \text{ for } i \neq j.$$

Since K_i^* is a Hall subgroup of Z and each $X \in \mathcal{E}^1(Z)$ lies in some K_i^*,

$$Z = K_0^* \times K_1^* \times \ldots \times K_n^*$$

and

$$K_i = \prod_{j \neq i} K_j^* \qquad (i = 0, 1, \ldots, n).$$

Furthermore, K_i^* is a Hall $\sigma(M_i)$-subgroup of Z, the subgroups M_i are pairwise not conjugate in G, and, for every element $z \in Z$, the factorization $z = \prod_{i=0}^{n} x_i$ with $x_i \in K_i^*$ is the σ-decomposition of z.

Define

$$T = Z - \bigcup_{i=0}^{n} K_i^*$$

and note that $t \in Z$ lies in T if and only if $z = yy'$ with $y \in K_i^{*\#}$ and $y \in K_i^{\#}$ for some index i. Therefore, by Lemma 14.6,

$$T \cap \tilde{H} \text{ is empty for each } H \in \mathcal{M}.$$

In particular, $\mathcal{C}_G(T)$ is disjoint from each of the sets $\mathcal{C}_G(\widetilde{M}_i)$.

Suppose $t \in T$, $g \in G$, and $t^g \in Z$. With y and y' as above, $y^g \in K_i^*$ and $y'^g \in K_i\#$ because $Z = K_i \times K_i^*$ and the orders of K_i and K_i^* are relatively prime. Therefore, by Proposition 14.2(d), $g \in K_i \times K_i^*$.

This proves that T is a TI-subset of G with $N_G(T) = Z$. Consequently, with $k_i = |K_i|$, $k_i^* = |K_i^*|$, and $z = |Z| = k_i k_i^*$,

$$|\mathcal{C}_G(T)| = |T||G : N_G(T)| = |T||G : Z|$$

$$= \left(z + n - \sum_{i=0}^{n} k_i^* \right) |G : Z| = \left(1 + \frac{n}{z} - \sum_{i=0}^{n} \frac{1}{k_i} \right) |G|.$$

Suppose all the M_i lie in $\mathscr{M}_{\mathscr{P}_1}$, which means that K_i always complements $M_{i\sigma}$ in M_i. Then, by Lemma 14.5,

$$|\mathscr{C}_G(\widetilde{M_i})| = (|M_{i\sigma}| - 1)|G : M_i|$$

$$= \left(\frac{1}{k_i} - \frac{1}{|M_i|}\right)|G| \geq \left(\frac{1}{k_i} - \frac{1}{2z}\right)|G|,$$

and the sets $\mathscr{C}_G(\widetilde{M_i})$ are pairwise disjoint. Now

$$|G^\#| \geq |\mathscr{C}_G(T)| + \sum_{i=0}^{n} |\mathscr{C}_G(\widetilde{M_i})|$$

$$\geq \left(1 + \frac{n}{z} - \sum_{i=0}^{n} \frac{1}{k_i}\right)|G| - \frac{n+1}{2z}|G| + \left(\sum_{i=0}^{n} \frac{1}{k_i}\right)|G|$$

$$= \left(1 + \frac{2n - (n+1)}{2z}\right)|G| \geq |G|,$$

and this contradiction proves that some M_i is of type $\mathscr{P}_2$.

For M_i of type $\mathscr{P}_2$, by Proposition 14.2(g), $K_i = \prod_{j \neq i} K_j^*$ has prime order and $M_{i\sigma}$ is nilpotent. Therefore $K_i = K_j^*$ for $j \neq i$, and $n = 1$. Furthermore, $Z = K_i \times K_i^* = K_j^* \times K_j$ is cyclic because $K_i^* \subseteq M_{i\sigma}$ and $r(K_i^*) = r(K_j) = 1$.

For our TI-subset T we now have $T = Z - (K \cup K^*) = \widehat{Z}$ and

$$|\mathscr{C}_G(T)| = \left(1 - \frac{1}{k_0} - \frac{1}{k_1} + \frac{1}{z}\right)|G|$$

$$= \left(1 - \frac{1}{k} - \frac{1}{k^*} + \frac{1}{kk^*}\right)|G| = \left(1 - \frac{1}{k}\right)\left(1 - \frac{1}{k^*}\right)|G|$$

$$\geq \left(1 - \frac{1}{3}\right)\left(1 - \frac{1}{5}\right)|G| = \frac{8}{15}|G| > \frac{1}{2}|G|.$$

Furthermore, Proposition 14.2(d) implies that $\widehat{Z} \cap M^g$ is empty for all $g \in G - M$.

Suppose $H \in \mathscr{M}_{\mathscr{P}}$. Let L be a Hall $\kappa(H)$-subgroup of H, $L^* = C_{H_\sigma}(L)$, and $S = L \times L^* - (L \cup L^*)$. Then we also have

$$|\mathscr{C}_G(S)| > \frac{1}{2}|G|,$$

and it follows that $\mathscr{C}_G(T) \cap \mathscr{C}_G(S)$ is not empty.

We want to prove that H is conjugate to M_0 or M_1 in G, and for this we can assume that $T \cap S$ is not empty. Then $L^* \cap K_i^* \neq 1$ for some i, and, by Proposition 14.2(c), $Y \in \mathcal{E}^1(L^* \cap K_i^*)$ satisfies

$$\{H\} = \mathscr{M}(C_G(Y)) = \{M_i\},$$

as desired.

Since $M_\sigma \subseteq M'$, the normal complement UM_σ of K in M, obtained in Proposition 14.2(a), is contained in M', and, since K is cyclic, UM_σ must be equal to M'. Then, by definition of $\tau_1(M)$, K is a Hall $\tau_1(M)$-subgroup of M. Thus $\kappa(M) = \tau_1(M)$.

Next let Y be a Hall $\sigma(M)$-subgroup of M_1 that contains K^*. By Proposition 14.2(f), $Y \subseteq M_\sigma$. Then $[Y, X_1] \subseteq M_\sigma \cap M_{1\sigma} = 1$. Therefore $Y \subseteq C_{M_\sigma}(X_1) = K^*$. It follows that $K^* = M_\sigma \cap M_1 \lhd M \cap M \cap M_1$, and, by Proposition 14.2(b1) applied to M_1 and K^* in place of M and K, $N_{M_1}(K^*) = K^* \times K$. Thus $M \cap M_1 = K^* \times K = Z$, and this completes the proof of the theorem. $\square$

Corollary 14.8. The maximal subgroups in $\mathscr{M}_{\mathscr{P}_1}$, if any, are all conjugate in G and, if $\mathscr{M}_{\mathscr{P}}$ is not empty, then $\mathscr{M}_{\mathscr{P}}$ contains exactly two classes of maximal subgroups conjugate in G.

Proof. This follows directly from Theorem 14.7(f) and (g). $\square$

Corollary 14.9. Choose $M_1, \ldots, M_n \in \mathscr{M}$ in such a way that each subgroup $H \in \mathscr{M}$ is conjugate in G to exactly one M_i.

(a) If $\mathscr{M}_{\mathscr{P}}$ is empty, then $G^\#$ is the disjoint union of the sets $\mathscr{C}_G(\widetilde{M_i})$, $i = 1, \ldots, n$.

(b) If $\mathscr{M}_{\mathscr{P}}$ is not empty, then, with $\widehat{Z}$ as in Theorem 14.7(e), $G^\#$ is the disjoint union of $\mathscr{C}_G(\widehat{Z})$ and the sets $\mathscr{C}_G(\widetilde{M_i})$, $i = 1, \ldots, n$.

Proof. By Lemma 14.5(b), the sets $\mathscr{C}(\widetilde{M_i})$ are pairwise disjoint. By definition of $\widetilde{M}$ for $M \in \mathscr{M}$, their union is the set $\widetilde{G}$ of all elements of xx' with $\ell_\sigma(x) = 1$ and $x' \in R(x)$.

If $\mathscr{M}_{\mathscr{P}}$ is empty, which means that $\kappa(M)$ is empty for each $M \in \mathscr{M}$, then case (2) of Lemma 14.6 does not occur, and it follows that $G^\# = \widetilde{G}$. This proves (a).

For the proof of (b), assume the notation of Theorem 14.7. Clearly, every $g \in \widehat{Z}$ satisfies condition (2) of Lemma 14.6, and it remains to show that every element $g = yy'$, with $y \in H_\sigma \#$ for some $H \in \mathscr{M}$ and y' a nonidentity $\kappa(M)$-element of $C_H(y)$, is conjugate to an element of $\widehat{Z}$. Since $\kappa(H)$ is not empty, $H \in \mathscr{M}_{\mathscr{P}}$ and H is conjugate to M or M^* by Theorem 14.7(g). So we may assume that $H = M$. Then y' is conjugate in M to an element of K. We may assume $y' \in K$, and then $y \in C_{M_\sigma}(y') = K^*$ by Theorem 14.7(d). Thus $g \in \widehat{Z}$. $\square$

Corollary 14.10. For every element $g \in G$ we have $\ell_\sigma(g) \leq 2$.

Proof. If $g \in \widetilde{M}$ with $M \in \mathscr{M}$, then $\ell_\sigma(g) \leq 2$ by definition of $\widetilde{M}$. If $t \in Z = K \times K^*$ in the situation of Theorem 14.7, then $\ell_\sigma(t) = 2$ because $K^* \subseteq M_\sigma$ and $K \subseteq M_\sigma^*$. Now apply Corollary 14.9. $\square$

The next lemma might be of independent interest, but we need it here only as a convenient tool for the investigation of certain maximal subgroups related to those of type $\mathscr{P}_2$.

Lemma 14.11. Suppose that $M \in \mathscr{M}_{\mathscr{F}}$, E is a complement of M_σ in M, $q \in \pi(E)$, $Q \in \mathcal{E}_q{}^1(E)$, and $Q \not\subseteq F(E)$. Then there exists $M^* \in \mathscr{M}$ such that either

(1) $q \in \tau_2(M^*)$ and $\mathscr{M}(C_G(Q)) = \{M^*\}$, or
(2) $q \in \kappa(M^*)$ and $M^* \in \mathscr{M}_{\mathscr{P}_1}$.

Actually, we have the situation of both Theorem 12.7 and Corollary 12.9, with $A_0 = [E, Q]$, $E = C_E(Q)A_0$, and $M^* \in \mathscr{M}(N_G(A))$.

Proof. By Corollary 12.6(a), every $A \in \mathcal{E}^2(E)$ is normal in E and thus contained in $F(E)$. Hence $q \notin \tau_2(M)$. By Lemma 12.1(b) and (d), E has a cyclic normal Hall $\tau_3(M)$-subgroup. Hence $q \notin \tau_3(M)$.

This shows that $q \in \tau_1(M)$. In particular, E has cyclic Sylow q-subgroups. Consequently, since $Q \not\subseteq F(E)$, $F(E)$ is a q'-group. By Corollary 12.10(b), E' is abelian.

It follows that $K = [E, Q]$ is an abelian q'-group. Therefore, because $KQ \lhd E$, the Frattini argument yields $E = KN_E(Q) = KC_E(Q)$. Hence $K = [E, Q] = [K, Q]$.

Now it follows from Proposition 10.11(d) that $[K, Q] = K$ is a cyclic normal subgroup of M. By Lemma 10.5, this implies that $\pi(K) \subseteq \tau_2(M)$. Take $p \in \tau_2(M)$ and $A \in \mathcal{E}_p{}^2(E)$. By Lemma 12.8(e), because $Q \not\subseteq Z(E)$, we have the situation of Theorem 12.7 with $A_0 = C_A(M_\sigma) = C_E(M_\sigma)$ equal to K.

Take $M^* \in \mathscr{M}(N_G(A))$. By Lemma 12.11, $p \in \sigma(M^*) - \beta(M^*)$ and q lies in $\tau_1(M^*)$ or $\tau_2(M^*)$ because $Q \not\subseteq C_E(A)$.

If $q \in \tau_1(M^*)$, then $q \in \kappa(M^*)$ because $C_{M^*}(Q) \supseteq C_A(Q) \supset 1$, and if $q \in \tau_2(M^*)$, then, for the same reason, Corollary 12.10(e) yields $\mathscr{M}(C_G(Q)) = \{M^*\}$.

In case $q \in \kappa(M^*)$, $M^* \in \mathscr{M}_{\mathscr{P}}$. Then we have $M^* \in \mathscr{M}_{\mathscr{P}_1}$ by Proposition 14.2(g) because $\sigma(M^*) \neq \beta(M^*)$. $\square$

Corollary 14.12. Suppose $M \subset \mathscr{M}_{\mathscr{P}_2}$. Let K, M^*, and K^* be as in Theorem 14.7 and U as in Proposition 14.2(a). Suppose $r \in \pi(U)$ and R is the Sylow r-subgroup of the abelian group U. Take $H \in \mathscr{M}(N_G(R))$.

Then $H \in \mathscr{M}_{\mathscr{F}}$, $U \subseteq H_\sigma$, $M \cap H = UK$, $N_H(U) \not\subseteq M$, $K \subseteq F(H \cap M^*)$, and $H \cap M^*$ is a complement of H_σ in H.

Proof. If H is conjugate to M in G, then R is a Sylow r-subgroup of H, and it follows that $r \in \sigma(H)$ because $N_G(R) \subseteq H$. But $r \neq \sigma(M)$, and hence H is not conjugate to M in G.

By Proposition 14.2(d), applied to M^* and K in place of M and K^*, M^* is the only conjugate of M^* in G that contains K. For some application below, note that this forces every subgroup of G in which K is subnormal to lie in M^*.

Now if H is conjugate to M^*, then $U \subseteq H = M^*$, contrary to the fact that $M \cap M^* = K \times K^*$ by Theorem 14.7(d). Since every element of $\mathscr{M}_{\mathscr{P}}$ is conjugate to M or M^* by Theorem 14.7(g), this proves that $H \in \mathscr{M}_{\mathscr{F}}$.

By Theorem 13.9, $H_\sigma \cap M_\sigma^* = 1$. Hence K lies in some complement D of H_σ in H. By Proposition 14.2(g), K has prime order, say q.

Suppose $K \nsubseteq F(D)$. Then Lemma 14.11 yields some $H^* \in \mathscr{M}$ such that either $q \in \tau_2(H^*)$ and $\mathscr{M}(C_G(K)) = \{H^*\}$ or $q \in \kappa(H^*)$ and $H^* \in \mathscr{M}_{\mathscr{P}_1}$. Then H^* is not conjugate to M^* in G because $q \in \sigma(M^*)$. Therefore $H^* \notin \mathscr{M}_{\mathscr{P}_1}$ by Theorem 14.7(g) since $M \in \mathscr{M}_{\mathscr{P}_2}$. On the other hand, M^* lies in $\mathscr{M}(C_G(K))$. This contradiction shows that $K \subseteq F(D)$.

Now K is subnormal in D. Therefore $D \subseteq M^*$. Recall that U satisfies Proposition 14.2(a), so that $U = [U, K]C_U(K) = [U, K]$. Since $K \subseteq \mathcal{O}_q(D)H_\sigma \lhd H$, it follows that

$$U = [U, K] \subseteq \mathcal{O}_q(D)H_\sigma \cap U \subseteq H_\sigma.$$

By Lemma 14.1, $N_M(U) = UK$ and $H_\sigma \subseteq F(H)$. Consequently, because $U \subseteq H_\sigma$, we obtain $M \cap H = UK$ and $N_H(U) \nsubseteq M$.

Finally suppose $D \neq H \cap M^*$. Then $H_\sigma \cap M^* \neq 1$ and $[H_\sigma \cap M^*, K] \subseteq H_\sigma \cap M_\sigma^* = 1$ because $K \subseteq M_\sigma^*$. Thus $C_{H_\sigma}(K) \neq 1$, and this implies that $q \in \tau_2(H)$ because $\kappa(H)$ is empty. Then, by Theorem 12.5(e), $H_\sigma \cap M^* = 1$ because $D \subseteq M^*$ and $\mathcal{E}_q^2(D)$ is not empty. This contradiction completes the proof of the corollary. $\square$

As a last application of Theorem 14.7 in this section, we obtain some more information on the situation of Theorem 14.4.

Lemma 14.13. Assume the situation of Theorem 14.4 in the case for which $|\mathscr{M}_\sigma(x)| > 1$.

  (a) If $\sigma(N)$ is not disjoint from $\pi(M)$, then $M \in \mathscr{M}_{\mathscr{F}}$ and $\tau_2(M)$ is empty, i.e., M is a Frobenius group with kernel M_σ.
  (b) If $y \in M_\sigma{}^{\#}$, $C_G(y) \nsubseteq M$, $g \in G$, and $N(y)^g = N$, then M contains an element m such that $N(y)^m = N$.

Proof. (a) Take $q \in \pi(M) \cap \sigma(N)$, $Q \in \mathcal{E}_q^1(M)$, and $M^* \in \mathscr{M}(N_G(Q))$. Since $\sigma(M)$ is disjoint from $\sigma(N)$ by Theorem 13.9, Q lies in some complement E of M_σ in M. By Theorem 14.4(d),

$$q \in \pi(M) \cap \sigma(N) \subseteq \beta(N) \subseteq \beta(G).$$

For some $g \in G$, $Q \subseteq N^g$. By Corollary 12.14, $M^* = N^g$ because $q \in \beta(N^g)$. By Theorem 14.4(c) and Lemma 12.1(g),

$$\pi(\langle x \rangle) \subseteq \tau_2(N) \subseteq \sigma(M) - \beta(M).$$

In particular, $\sigma(M) \neq \beta(M)$ and $\sigma(M \cap \pi(M^*)) \nsubseteq \kappa(M^*)$. Hence Proposition 14.2(g) yields $M \notin \mathscr{M}_{\mathscr{P}_2}$. If $M \in \mathscr{M}_{\mathscr{P}_1}$, then $q \in \kappa(M)$, and, by

Theorem 14.7(a) and (b), $\sigma(M) \cap \pi(M^*) \subseteq \kappa(M^*)$, which is false. Thus $M \notin \mathcal{M}_{\mathcal{P}_1}$, whence $M \in \mathcal{M}_{\mathcal{F}}$.

Suppose $\tau_2(M)$ is not empty. Take any $p \in \tau_2(M)$. By Lemma 12.1(g), $p \notin \beta(G)$. Hence the inclusions above show that $p \notin \sigma(N) \cup \tau_2(N)$. It then follows that all the elements of $\mathcal{E}_p{}^1(N)$ are conjugate in N because $r_p(N) \leq 1$.

Now take $A \in \mathcal{E}_p{}^2(E)$. Since $C_G(Q) \subseteq N^g$, we have $[A, Q] \neq 1$. By Corollary 12.10(c), $q \in \tau_1(M)$. Then Corollary 12.9 yields subgroups A_0, $A_1 \in \mathcal{E}_p{}^1(A)$, not conjugate in G, such that

$$A_0 \subseteq C_G(M_\sigma) \subseteq C_G(x) \subseteq N \text{ and } A_1 \subseteq C_G(Q) \subseteq N^g.$$

Obviously this contradicts the above statement about $\mathcal{E}_p{}^1(N)$. Therefore $\tau_2(M)$ is empty and the proof of (a) is complete.

(b) Since $C_G(y) \not\subseteq M$, Theorem 14.4 shows that $|\mathcal{M}_\sigma(y)| > 1$ and then that $N(y)$ is defined. By Theorem 14.4, both $M \cap N$ and $(M \cap N(y))^g$ complement N_σ in N. Hence there exists $n \in N$ such that $(M \cap N(y))^{gn} = M \cap N$. Then $x \in M^{gn}$ and Theorem 14.4 yields an element $c \in C_N(x)$ such that $M^{gnc} = M$.

Now $m = gnc$ lies in $N_G(M) = M$ and satisfies $N(y)^m = N(y)^{gnc} = N^{nc} = N$. $\square$

15. The Subgroup M_F

In this section we will prove some results about the Fitting subgroup $F(M)$ and the largest normal nilpotent Hall subgroup M_F of a maximal subgroup M of G. It is easy to see that M_F is the product of the normal Sylow subgroups of M and so is well defined and lies in M_σ.

Throughout this section we use the following notation:

M is a maximal subgroup of G,

K is a Hall $\kappa(M)$-subgroup of M,

U is a complement of KM_σ in M for which

KU is a complement to M_σ in M.

Lemma 15.1. The following conditions hold.

(a) $UM_\sigma \triangleleft M = KUM_\sigma$, K is cyclic, $M_\sigma \subseteq M'$, and M'/M_σ is abelian.

(b) If $K \neq 1$, then $M' = UM_\sigma$ and U is abelian.

(c) If X is a nonidentity subgroup of U such that $C_{M_\sigma}(X) \neq 1$, then $\mathcal{M}(C_G(X)) = \{M\}$ and X is a cyclic $\tau_2(M)$-subgroup.

(d) The group $\langle C_U(x) \mid x \in M_\sigma^\# \rangle$ is abelian.

(e) If $U \neq 1$, then U contains a subgroup U_0, of the same exponent as U, such that $U_0 M_\sigma$ is a Frobenius group with kernel M_σ.

Proof. By Theorem 14.7(d) and (h), K is cyclic and if $K \neq 1$, then $M' = UM_\sigma$. By Corollary 12.10(b), $(M/M_\sigma)'$ is abelian and, by Theorem 10.2(c), $M_\sigma \subseteq M'$. This proves (a) and (b).

By Corollary 14.3, the group X in (c) must be a $\tau_2(M)$-group and we have $\mathscr{M}(C_G(X)) = \{M\}$ if X is cyclic. Since M has an abelian Hall $\tau_2(M)$-subgroup by Corollary 12.10(b) and, because $C_{M_\sigma}(A) = 1$ for every $A \in \mathscr{E}_p^2(U)$ by Theorem 12.5(d), X is indeed cyclic.

In Theorem 12.12, (d) and (e) have been proved under the assumption that $\kappa(M)$ is empty, i.e., $K = 1$. If $K \neq 1$, then U is abelian by (b). In this case (d) is trivial and (e) is obvious from the very short and easy argument in the proof of Theorem 12.12 that deals with the case that $C_E(S) = E$. □

Theorem 15.2. For every $M \in \mathscr{M}$, $1 \subset M_F \subseteq M_\sigma \subseteq M' \subset M$. Suppose $M_F \neq M_\sigma$ and let $p = |K|$, $K^* = C_{M_\sigma}(K)$, and $q = |K^*|$. Then

(a) M is of type $\mathscr{P}_1$, i.e., $M = KM_\sigma$,

(b) p and q are primes and $q \in \pi(M_F) \cap \beta(M)$,

(c) M has a normal Sylow q-subgroup Q (contained in M_F),

(d) a complement D of Q in $M_\sigma = M'$ is nilpotent,

(e) $Q_0 = C_Q(D)$ is a normal subgroup of M,

(f) $\overline{Q} = Q/Q_0$ is a minimal normal subgroup of M/Q_0 and is elementary abelian of order q^p, and

(g) $M'' = M_\sigma' \subseteq F(M) = QC_M(Q) = C_M(\overline{Q}) = C_{M_\sigma}(\overline{K^*}) \subset M_\sigma$, and $M_\sigma = M'$.

Proof. As mentioned earlier in this section, $M_F \subseteq M_\sigma$ and $M_\sigma \subseteq M'$. Clearly $1 \subset M_\sigma$ and $M' \subset M$. We must show that $M_F \neq 1$. So assume that $M_F \subset M_\sigma$, i.e., M_σ is not nilpotent. By Lemma 14.1, this implies (a) and, by Theorem 14.7(f), (a) implies that $q = |K^*|$ is a prime.

Let K_1 be a subgroup of K having prime order. By Proposition 14.2(a), K acts in a prime manner on M_σ. By Lemma 6.3(a), $M_\sigma = [M_\sigma, K]$. Thus $[M_\sigma, K] \not\subseteq F(M_\sigma)$. Therefore, by Theorem 3.8, $K^* \cap F(M) \neq 1$. Thus K^* lies in $Q = \mathcal{O}_q(M)$. It follows from Proposition 1.5(d) that K acts regularly on M_σ/Q. Therefore, by Theorem 3.7 (applied to the group $K_1 M_\sigma/Q$), M_σ/Q is nilpotent. This proves (c) and (d).

By Proposition 1.5(a), we may choose a K-invariant complement D of Q in M_σ. Then $Q_0 = C_Q(D)$ is a KD-invariant proper subgroup of Q because M_σ is not nilpotent. Then $N_Q(Q_0) \supset Q_0$ and $N_M(Q_0)/Q_0$ has a minimal normal subgroup Q_1/Q_0 such that $Q_1 \subseteq Q$. If $K^* \subseteq Q_0$ or $K^* \not\subseteq Q_1$, then K acts regularly on DQ_1/Q_0, and therefore Theorem 3.7 implies that D centralizes Q_1/Q_0. But, by Proposition 1.5, this implies that $Q_1 = C_{Q_1}(D)Q_0 = Q_0$, a contradiction. This proves that $Q_1 \supseteq K^*$ and $Q_0 \not\supseteq K^*$. The same argument, with Q_1 in place of Q_0, would show that $Q_1 \not\supseteq K^*$ if $Q_1 \subset Q$. Thus $Q_1 = Q$. This proves (e) and the first two assertions of (f).

Since D is nilpotent, but M_σ is not, Proposition 1.5(d) yields

$$F(M) = QC_M(Q) = C_{M_\sigma}(\overline{Q}) \subset M_\sigma.$$

In particular, $C_{\overline{Q}}(D) \neq \overline{Q}$. Therefore, by minimality of $\overline{Q}$, $C_{\overline{Q}}(D) = 1$. Hence we can apply Theorem 3.10 to the action of the Frobenius group KD on $\overline{Q}$ and deduce that $p = |K|$ is a prime, $|\overline{Q}| = q^p$, and $D' \subseteq C_D(\overline{Q})$. It follows that $M_\sigma' \subseteq QD' \subseteq C_{M_\sigma}(\overline{Q}) \subseteq C_{M_\sigma}(\overline{K^*})$, whence the latter subgroup is normal in M_σ and therefore in $KM_\sigma = M$. Then minimality of $\overline{Q}$ forces $C_{M_\sigma}(\overline{K^*})$ to centralize $\overline{Q}$ and this completes the proof of (g).

Finally, if $q \notin \beta(M)$, then Theorem 5.5(a) shows that $(DK)' = D$ centralizes Q, a contradiction. Thus $q \in \beta(M)$ and the proof is complete. $\square$

Corollary 15.3. Suppose H is a nonidentity Hall subgroup of M_σ. Then

(a) $C_M(H) = C_{M_\sigma}(H)X$ with X a cyclic $\tau_2(M)$-subgroup, and
(b) any two elements of H conjugate in G are already conjugate in $N_M(H)$.

Proof. By Proposition 14.2(b1) and (e), $C_M(H)$ is a $\kappa(M)'$-group. Hence $C_M(H) = C_{M_\sigma}(H)X$ with X conjugate in M to a subgroup of U and Lemma 15.1(c) yields (a).

Suppose $x, y \in H$, $g \in G$, and $x = y^g$. Then $x \in M^g$, and Theorem 14.4 yields an element $c \in C_G(x)$ such that $M^{gc} = M$. Then $gc \in M$ and $y^{gc} = x$.

So assume that H is not normal in M. Then M_σ is not nilpotent and, with Q as in Theorem 15.2, we have $QH \lhd M$ because M_σ/Q is nilpotent. It follows that $Q \cap H = 1$ and $M = N_M(H)Q$ by the Frattini argument. Write $gc = na$ with $n \in N_M(H)$ and $a \in Q$. Then $y^{na} = x$ and $axa^{-1}x^{-1} = y^nx^{-1} \in Q \cap H = 1$ because $Q \lhd M$. Thus $a \in C_M(x)$ and $y^n = x$ as required. $\square$

Corollary 15.4. Suppose H is a nonidentity nilpotent Hall subgroup of G. Then our maximal subgroup M can be chosen in such a way that $H \subseteq M_\sigma$.

Proof. Let S be a nonidentity Sylow subgroup of H and $M \in \mathcal{M}(N_G(S))$. Then $S \subseteq M_\sigma$ and Corollary 15.3(a), applied with S in place of H, shows that M_σ contains every Sylow subgroup of M that lies in $C_M(S)$. Thus $H \subseteq M_\sigma$ because H is nilpotent. $\square$

Corollary 15.5. Let $H = M_F$ and $Y = \mathcal{O}_{\sigma(M)'}(F(M))$. Then

(a) Y is a cyclic $\tau_2(M)$-subgroup of $F(M)$,
(b) $M'' \subseteq F(M) = C_M(H)H = F(M_\sigma) \times Y$,
(c) $H \subseteq M'$ and M'/H is nilpotent, and
(d) if $K \neq 1$, then $F(M) \subseteq M'$.

Proof. Parts (a), (b), and (c) follow directly from Lemma 15.1(a) if $H = M_\sigma$ and from Theorem 15.2(g) and Corollary 15.3(a) if $H \neq M_\sigma$. If $K \neq 1$, then $M_\sigma \subseteq M'$ and $M/M' \cong K$ by Lemma 15.1(a) and (b). Since K is a $\tau_2(M)'$-group by definition (see Section 14), M' contains Y. Thus (d) follows. $\square$

Corollary 15.6. Suppose $M \in \mathcal{M}_{\mathscr{P}}$ (i.e., $K \neq 1$). Then $K^* = C_{M_\sigma}(K)$ is a nonidentity cyclic subgroup of both M_F and M''. Furthermore, M_F is not cyclic.

Proof. By Theorem 14.7(h), $M = KM'$ and $K \cap M' = 1$. Therefore, by Lemma 6.3, $K^* \subseteq M''$. By Proposition 14.2(c) and Theorem 14.7(d), $K^* \neq 1$ and K^* is cyclic. By Theorem 15.2(b) and (c), $K^* \subseteq M_F$.

Finally, if M_F is cyclic, then $F(M)$ is cyclic by Corollary 15.5, and we have $M' \subseteq C_M(F(M)) \subseteq F(M)$ and $K^* \subseteq M'' = 1$, a contradiction. $\square$

Theorem 15.7. Suppose $F(M)$ is not a TI-subgroup of G. Let $H = M_F$ and choose $g \in G - M$ such that $X = F(M) \cap F(M)^g$ is not trivial. Take E, E_1, E_2, E_3 as in Sections 12-13. Then

(a) $M \in \mathcal{M}_{\mathscr{F}} \cup \mathcal{M}_{\mathscr{P}_1}$ and $H = M_\sigma$,
(b) $X \subseteq H$ and X is cyclic,
(c) $M' = F(M) = M_\sigma \times \mathcal{O}_{\sigma(M)'}(F(M))$,
(d) $E_3 = 1$, $E_2 \lhd E$, and $E/E_2 \cong E_1$, which is cyclic, and
(e) one of the following conditions holds:
  (1) $M \in \mathcal{M}_{\mathscr{F}}$ and H is abelian of rank two,
  (2) $p = |X|$ is a prime in $\sigma(M) - \beta(M)$, $\mathcal{O}_p(H)$ is not abelian, $\mathcal{O}_{p'}(H)$ is cyclic, and the exponent of M/H divides $q - 1$ for every $q \in \pi(H)$,
  (3) $p = |X|$ is a prime in $\sigma(M) - \beta(M)$, $\mathcal{O}_{p'}(H)$ is cyclic, $\mathcal{O}_p(H)$ has order p^3 and is not abelian, $M \in \mathcal{M}_{\mathscr{P}_1}$, and $|M/H|$ divides $p + 1$.

Proof. Take $p \in \pi(X)$ and $X_1 \in \mathcal{E}_p{}^1(X)$. If $\mathcal{O}_p(M)$ is cyclic, then X_1 is normal in both M and M^g, which is impossible. Thus $\mathcal{O}_p(M)$ is not cyclic and, by Corollary 15.5, $p \in \sigma(M)$. Hence $X \subseteq M_\sigma$. Consequently, by Theorem 10.1(a) and Lemma 12.17, $C_G(X_1) \not\subseteq M$ and X is cyclic.

Since $\mathcal{M}(C_H(X_1)) \neq \{M\}$, Theorem 12.13 and the Uniqueness Theorem show that $C_H(X_1)$ is abelian of rank less than 3. Thus $\pi(H)$ is disjoint from $\beta(M)$ and, by Proposition 14.2(g) and Theorem 15.2(b), $M \notin \mathcal{M}_{\mathscr{P}_2}$ and $H = M_\sigma$. This proves (a). Since $X \subseteq M_\sigma$, we also obtain (b).

Now H is a Hall $\beta(M)'$-subgroup of itself, and hence is centralized by E' because of Lemma 12.19. By Corollary 12.6(d), $C_{M_\sigma}(x) = 1$ for every $x \in E_3^\#$. But $E_3 \subseteq E'$. Therefore $E_3 = 1$. Consequently (c) and (d) follow from Corollary 15.5 and Lemma 12.1.

Suppose H is abelian. If $M \in \mathcal{M}_{\mathscr{P}_1}$, i.e., $K \neq 1$ and $U = 1$, then Lemma 15.1(b) shows that $H = M_\sigma = M'$, contrary to Corollary 15.6, which implies that $M'' \neq 1$. Therefore $M \in \mathcal{M}_{\mathscr{F}}$ and condition (1) of (e) holds.

So assume that H is not abelian. Then $\mathcal{O}_{p'}(H)$ is abelian because $C_H(X_1)$ is abelian. Hence $P = \mathcal{O}_p(H)$ is not abelian and X is a p-group.

Let $Z_0 = \Omega_1(Z(P))$. Clearly $X_1 \neq Z_0$. Let $B = X_1 \times Z_0$. Now we know $B \in \mathcal{E}^2(P) \cap \mathcal{E}^*(P)$ because $C_H(X_1)$ has rank less than 3. Thus $|Z_0| = p$ and $Z(P)$ is cyclic. Moreover, by Lemma 10.13(b), $C_P(X_1) = C_P(B) = X_1 \times Z$ with Z cyclic. Thus $X = X_1$.

Since $P \in \mathcal{U}$ by Theorem 12.13, Corollary 9.2 shows that $\mathcal{O}_{p'}(H) \in \mathcal{U}$ if $r(\mathcal{O}_{p'}(H)) \geq 2$. However we have $\mathcal{O}_{p'}(H) \notin \mathcal{U}$ because $C_H(X) \notin \mathcal{U}$. Consequently the abelian group $\mathcal{O}_{p'}(H)$ must be cyclic. It follows that $Z(H)$ is cyclic. Therefore, if A is any subgroup of M that acts regularly on $Z(H)$, then $|A|$ divides $q - 1$ for all $q \in \pi(H)$.

It follows that condition (2) of (e) holds if $M \in \mathcal{M}_{\mathscr{F}}$ because, in that case, Lemma 15.1(e) yields a subgroup of the same exponent as M/H, which acts regularly on H.

So we may assume that $M \in \mathcal{M}_{\mathscr{P}_1}$. By Theorem 14.7(f) and Corollary 15.6, $K^* = C_H(K)$ has prime order and is contained in M''. Since $M = HK$, K is cyclic, and $\mathcal{O}_{p'}(H)$ is abelian, we have $K^* \subseteq M'' \subseteq P$.

By Proposition 14.2, $C_H(k) = K^*$ for every $k \in K^{\#}$. Thus, if (c2) is false, then $K^* = \Omega_1(Z(P))$ and $|K|$ does not divide $p - 1$. Consequently $Z(P) \subseteq C_H(K) = K^*$ and, by Theorem 5.5(b), $r(P) = 2$. Hence Corollary 10.7(b) shows that P has order p^3. Then $|K|$ divides $p + 1$ by Theorem 2.5. Thus condition (3) of (e) holds and the proof of the theorem is complete. $\square$

As a last application of Theorem 15.2, we obtain some more information on the situation of Corollary 14.12.

Theorem 15.8 (Feit-Thompson, 1991 [9]). Assume the situation of Corollary 14.12 and suppose $\tau_2(H)$ is not empty. Let $q \in \pi(K)$. Then $q = |K|$, q is the unique prime in $\tau_2(H)$, and $\tau_2(M)$ is empty.

Proof. By Theorem 14.7(f), $|K| = q$. Let $D = H \cap M^*$, a complement of H_σ in H. There exists A in $\mathcal{E}^2(D)$. By Lemma 12.1(g), $A \in \mathcal{E}^*(G)$. Since K lies in $F(D)$ by Corollary 14.12, Corollary 12.6(a) yields

$$(15.1) \qquad A \subseteq C_G(K) \subseteq M^*.$$

As $q \in \sigma(M^*)$, Corollary 12.6(b) shows that $A \subseteq M^*_\sigma$. If $M^*_\sigma \neq M^*_F$, then

$$(15.2) \qquad F(M^*) \text{ contains } A \text{ and a Sylow } q\text{-subgroup } Q \text{ of } M^*$$

by (g) and (c) of Theorem 15.2. Since (15.2) is obvious if $M^*_\sigma = M^*_F$, it is always valid.

By (15.2), A centralizes Q if A is not a q-group, and then Corollary 9.2 shows that $Q \notin \mathcal{U}$ because $A \notin \mathcal{U}$. If A is a q-group, $A \in \mathcal{E}_q^*(G)$. In either case, it now follows from the Uniqueness Theorem (Theorem 9.6) that $q \notin \beta(G)$. Therefore, by Proposition 14.2(g) and Theorem 15.2(b), M^* is of type $\mathscr{P}_1$ and M^*_σ is nilpotent. Since, by Lemma 12.17,

$$K \subseteq C_{M^*_\sigma}(K^*) \subseteq M^{*'}_\sigma,$$

Q is not abelian. So $Q \in \mathscr{U}$ by Theorem 12.13. As noted above, this forces A to be a q-group. Since A was chosen arbitrarily in $\mathcal{E}^2(D)$,

$$\tau_2(H) = \{q\}.$$

Now, by Theorem 12.7(b), A contains a subgroup X of order q that centralizes H_σ. If $X = K$, then, by (15.1),

$$H = H_\sigma(H \cap M^*) \subseteq \langle C_G(K), M^* \rangle \subseteq M^*,$$

a contradiction. Thus $X \neq K$. As K is a Sylow q-subgroup of M, we have $X \not\subseteq M$. For U as in Corollary 14.12, UK is a complement to M_σ in M and $U \subseteq H_\sigma$. Therefore U centralizes X and

$$C_G(U) \not\subseteq M.$$

If there exists $r \in \tau_2(M)$, then Corollary 12.6(b) (applied to $\Omega_1(\mathcal{O}_r(U))$ in place of A) yields

$$C_G(U) \subseteq C_G(\Omega_1(\mathcal{O}_r(U))) \subseteq M,$$

a contradiction. Thus $\tau_2(M)$ is empty, as desired. $\square$

Corollary 15.9. Let $x \in M_\sigma{}^\#$ and $N \in \mathscr{M}(C_G(x))$. Assume that

$$C_G(x) \not\subseteq M \text{ and } N \notin \mathscr{M}_{\mathscr{F}}.$$

Take $r \in \pi(\langle x \rangle)$ and x_r of order r in $\langle x \rangle$. Then, for a suitable choice of a complement E to M_σ in M,

    (a) **(D. Sibley [24])** $M \in \mathscr{M}_{\mathscr{F}}$ and $N \in \mathscr{M}_{\mathscr{P}_2}$,
    (b) **(Feit and Thompson, 1991 [9])** E is cyclic and M is a Frobenius
         group, and
    (c) $r \in \tau_2(N)$, $N_E(\langle x_r \rangle) \subseteq E \cap N$, and $|E \cap N| = |N/N'|$.

Proof. Take $y \in C_G(x) - M$. Then $M, M^y \in \mathscr{M}_\sigma(x)$ and $M \neq M^y$. Hence we are in the situation of Theorem 14.4 with $|\mathscr{M}_\sigma(x)| \geq 2$. Therefore $C_{N_\sigma}(x) \neq 1$ and

(15.3)
$$\{N\} = \mathscr{M}(C_G(x)), \ N \in \mathscr{M}_{\mathscr{P}_2}, \ r \in \tau_2(N) \cap \sigma(M),$$
$$\text{and } M \cap N \text{ is a complement to } N_\sigma \text{ in } N.$$

Let K_1 be a Hall $\kappa(N)$-subgroup of $M \cap N$. By Proposition 14.2(g) and (a), $|K_1|$ is prime and there exists an abelian normal complement U_1 to K_1 in $M \cap N$ for which

(15.4) $$\hspace{3cm} C_{U_1}(K_1) = 1.$$

Let R be a Sylow r-subgroup of $M \cap N$ (and therefore also of N). Since $r \in \tau_2(N) \cap \sigma(M)$,

$$(15.5) \qquad\qquad R \subseteq U_1; \text{ also } N_G(R) \subseteq M,$$

by Corollary 12.10(d). Moreover, by Corollary 14.12, with N and M in place of M and H, respectively, $M \in \mathcal{M}_{\mathcal{F}}$. This proves (a) and shows that M_σ is nilpotent. Hence $M_\sigma \subseteq F(M)$. Since $K_1 R$ is not nilpotent, by (15.2), $K_1 \not\subseteq M_\sigma$. Thus $K_1 \cap M_\sigma = 1$. We choose E to contain K_1.

Now $x \in M_\sigma \subseteq F(M)$ and $C_G(x) \not\subseteq M$. Hence $F(M)$ is not a TI-subgroup of G. Since $\tau_2(N)$ is not empty, $\tau_2(M)$ is empty by Theorem 15.8. Thus $E_2 = 1$. By Theorem 15.7(d), E is cyclic. Moreover, $\kappa(M)$ is empty because $M \in \mathcal{M}_{\mathcal{F}}$. Therefore M is a Frobenius group and (b) follows.

As $N_G(\langle x_r \rangle) \supseteq C_G(x)$, we have $N_G(\langle x_r \rangle) \subseteq N$ by (15.1). Consequently $E \cap N \supseteq N_E(\langle x_r \rangle)$. Since

$$K_1 \subseteq E \cap N \subseteq M \cap N = K_1 U_1 \text{ and } C_{U_1}(K_1) = 1,$$

$K_1 = C_{E \cap N}(K_1)$. But then $E \cap N = K$ because E is cyclic. Now (c) follows from (15.1). $\square$

16. The Main Results

Here we obtain our main results on the structure and embedding of the maximal subgroups of G. These are Theorems A–E and, more or less equivalently, Theorems I and II. We state them together with the prerequisite definitions and notation not given in Section 1. The latter allow the reader to compare conveniently what we have achieved here with the main results of Chapter IV in **FT**, Theorems 14.1 and 14.2. As mentioned earlier, they are best understood as generalizations of intermediate theorems in the proof of the Feit-Hall-Thompson CN-theorem. Specifically, Theorem I is analogous to Theorem 14.1.5 of **G**, which asserts that a nonnilpotent solvable CN-group is either a Frobenius group or a three step group (as defined in **G**, p. 401, not as defined here). Theorem II is analogous to Theorem 14.2.3 of **G**, which asserts that in a minimal counterexample G to the CN-theorem every maximal nilpotent subgroup H is disjoint from its conjugates (so it is a trivial intersection set in G with normalizer $N_G(H)$).

Like the analogous results in the proof of the CN-theorem our main results lay the basis for the character theoretic part of the proof.

Throughout this section let M denote an arbitrary maximal subgroup of G and let M_F denote the largest nilpotent normal Hall subgroup of M. In **FT** the group M_F somehow plays the role of the Frobenius kernel of M and is used in the statement of the main results in an essential way. So far in these notes most arguments and results have centered around a certain normal Hall subgroup M_σ of M (equal to M_F in the CN-case). Actually M_σ and M_F are closely related and nearly always coincide.

For the statement of the main results it is convenient to use the two prime sets $\sigma(M) \subseteq \pi(M)$ and $\kappa(M) \subseteq \pi(M) - \sigma(M)$. Although their exact definitions are not of importance for this purpose, we repeat them here for the sake of clarity:

$$\sigma(M) = \{p \in \pi(M) \mid N_G(P) \subseteq M \text{ for some}$$
$$\text{(and hence every) Sylow } p\text{-subgroup } P \text{ of } M\},$$
$$\kappa(M) = \{p \in \pi(M) - \sigma(M) \mid \text{ every Sylow } p\text{-subgroup } P \text{ of } M \text{ is cyclic}$$
$$\text{and contains a nonidentity element } x \text{ such that } C_{M_\sigma}(x) \neq 1\}.$$

For any subset X of G, $\mathscr{M}(X)$ denotes the set of all maximal subgroups of G that contain X. For arbitrary subsets X and Y of G, we have defined

$$\mathscr{C}_Y(X) = \{x^y \mid x \in X \text{ and } y \in Y\}.$$

We now state Theorems A–E and give their proof afterwards.

Theorem A. The following conditions are satisfied by M.

(1) M has a unique normal Hall $\sigma(M)$-subgroup M_σ and M_σ is also a $\sigma(M)$-Hall subgroup of G.
(2) M has a cyclic Hall $\kappa(M)$-subgroup K.
(3) KM_σ has a K-invariant complement U in M and thus

$$UM_\sigma \triangleleft M = KUM_\sigma \text{ and } U \triangleleft UK.$$

(4) $C_U(k) = 1$ for every $k \in K^\#$.
(5) $K^* = C_{M_\sigma}(K)$ is not trivial and if $K \neq 1$, then $C_M(k) = K \times K^*$ for every $k \in K^\#$.
(6) $1 \subset M_F \subseteq M_\sigma \subseteq M' \subset M$ and M'/M_F is nilpotent.
(7) $M'' \subseteq F(M) = C_M(M_F)M_F$ and if $K \neq 1$, then $F(M) \subseteq M'$.
(8) If $M_F \neq M_\sigma$, then $U = 1$, $F(M)$ is a TI-subset of G, and K has prime order.

For Theorems B–E we introduce some further notation:

$$Z = K \times K^*,$$
$$\widehat{Z} = Z - (K \cup K^*),$$
$$\widehat{M_\sigma} = \{a \in M \mid C_{M_\sigma}(a) \neq 1\},$$
$$A(M) = \widehat{M_\sigma} \cap UM_\sigma,$$
$$A_0(M) = \widehat{M_\sigma} - \mathscr{C}_M(K^\#).$$

Since $M_\sigma \subseteq M' \subset M = KUM_\sigma$, each of the groups K and U can be trivial, but not both. There are three cases. In accord with previous notation they are:

$$(\mathscr{F}) \quad K = 1 \text{ and } U \neq 1,$$
$$(\mathscr{P}_1) \quad K \neq 1 \text{ and } U = 1,$$
$$(\mathscr{P}_2) \quad K \neq 1 \text{ and } U \neq 1.$$

We have defined $\mathscr{M}_{\mathscr{P}}$ = the set of all maximal subgroups of type $\mathscr{P}_1$ or $\mathscr{P}_2$.

Theorem B. The following conditions are satisfied by M.

(1) Every Sylow subgroup of U is abelian of rank at most two.
(2) $\langle U \cap \widehat{M}_\sigma \rangle$ is abelian.
(3) U has a subgroup U_0 that has the same exponent as U and satisfies $U_0 \cap \widehat{M}_\sigma = 1$.
(4) $\mathscr{M}(C_G(X)) = \{M\}$ for every nonidentity subgroup X of U such that $C_{M_\sigma}(X) \neq 1$.
(5) The set $A(M) - M_\sigma$ is a TI-subset of G.

Theorem C. Suppose $K \neq 1$. Then the following conditions hold.

(1) U is abelian and $N_G(U) \not\subseteq M$.
(2) K^* is cyclic, $1 \subset K^* \subseteq M_F$, but M_F is not cyclic.
(3) $M' = UM_\sigma$ and $K^* \subseteq M''$.
(4) There exists a unique $M^* \in \mathscr{M}_{\mathscr{P}}$ such that $K = C_{M^*_\sigma}(K^*)$ and K^* is a Hall $\kappa(M^*)$-subgroup of M^*.
(5) $\mathscr{M}(C_G(X)) = \{M\}$ and $\mathscr{M}(C_G(Y)) = \{M^*\}$ for all subgroups $X \subseteq K^*$ and $Y \subseteq K$ of prime order.
(6) $M \cap M^* = Z = K \times K^*$, which is a cyclic group.
(7) M or M^* is of type $\mathscr{P}_2$ and every $H \in \mathscr{M}_{\mathscr{P}}$ is conjugate to M or M^* in G.
(8) $\widehat{Z}$ is a TI-subset of G with $N_G(\widehat{Z}) = Z$.
(9) $\mathscr{C}_M(\widehat{Z})$ is equal to $A_0(M) - A(M)$ and is a TI-subset of G.
(10) If $U \neq 1$, then K has prime order and $F(M)$ is a TI-subset of G containing M_σ.
(11) If $U = 1$, then K^* has prime order.

Theorem D. The following conditions are satisfied by M.

(1) Whenever two elements of M_σ are conjugate in G, they are conjugate in M.
(2) For every $g \in G - M$, the group $M_\sigma \cap M^g = M_\sigma \cap M_\sigma{}^g$ is cyclic.
(3) For every $x \in M_\sigma{}^\#$, $C_M(x)$ is a Hall subgroup of $C_G(x)$ and has a normal complement $R(x)$ in $C_G(x)$ that acts sharply transitively by conjugation on the set $\{ M^g \mid g \in G, x \in M^g \}$.
(4) If $x \in M_\sigma{}^\#$ and $C_G(x) \not\subseteq M$, then $C_G(x)$ lies in a unique maximal subgroup $N = N(x)$ of G. Furthermore,

$$R(x) = C_{N_\sigma}(x), \ N_\sigma = N_F, \ x \in A(N) - N_\sigma, \ N \in \mathscr{M}_{\mathscr{F}} \cup \mathscr{M}_{\mathscr{P}_2},$$

and $M \cap N$ is a complement of N_σ in N. Also, if $N \in \mathscr{M}_{\mathscr{P}_2}$, then $M \in \mathscr{M}_{\mathscr{F}}$, M is a Frobenius group with cyclic Frobenius complement, and M_F is not a TI-subgroup of G.

Theorem E. For each $x \in M_\sigma^{\#}$, take $R(x)$ as in Theorem D. Define

$$\widetilde{M} = \bigcup_{x \in M_\sigma^{\#}} xR(x).$$

Then

(1) $|\mathscr{C}_G(M)| = (|M_\sigma| - 1)|G : M|$.

Let $M_1, \ldots, M_n$ be maximal subgroups of G such that every maximal subgroup of G is conjugate in G to exactly one M_i. Then

(2) $\pi(G)$ is the disjoint union of the sets $\sigma(M_i)$.

Let $\widetilde{G}$ be the union of the sets $\mathscr{C}_G(\widetilde{M_i})$. Then

(3) $\widetilde{G}$ is the disjoint union of the sets $\mathscr{C}_G(\widetilde{M_i})$, $G^{\#} = \widehat{G}$ if $\mathscr{M}_{\mathscr{P}}$ is empty, and $G^{\#}$ is the disjoint union of $\widetilde{G}$ and $\mathscr{C}_G(\widehat{Z})$ if $\mathscr{M}_{\mathscr{P}}$ is not empty and $M \in \mathscr{M}_{\mathscr{P}}$.

The cosets $xR(x)$ appearing in Theorem E are denoted by A_x in **FT**. They are used in the study of so-called tamely imbedded subsets of G (defined in Section 13 of **FT** and later in this section). In Lemma 14.5(a) we have shown that $xR(x) \cap yR(y)$ is empty whenever x and y are distinct elements of G for which these sets are defined.

All these results have already appeared elsewhere in this book or are a direct consequence of previous results. Consequently their "proof" can be given schematically.

Theorem A

Theorem 10.2(b)	$\rightarrow$	(1)
Lemma 15.1(a)	$\rightarrow$	(2)
Proposition 14.2(a)(b)(c)	$\rightarrow$	(3)(4)(5)
Theorem 15.2(a)		
Corollary 15.5	$\rightarrow$	(6) (7)(8)
Theorem 15.7(a)(b)		

Theorem B

Lemma 12.1(d)
$\left.\right\}$ $\rightarrow$ (1)
Theorem 12.5(b)

Lemma 15.1(d)(e)(c) $\rightarrow$ (2)(3)(4)(5)

Theorem C

Corollary 14.12
Corollary 15.6 $\left.\right\}$ $\rightarrow$ (1) (2)(3)
Lemma 15.1(b)

Theorem 10.1(b)
Theorem 14.7(a)(b)(c) $\left.\right\}$ $\rightarrow$ (4) (5)
Proposition 14.2(c)

Theorem 14.7(d)(f)(g)(e) $\rightarrow$ (6)(7)(11)(8)

Proposition 14.2(d)
$\left.\right\}$ $\rightarrow$ (9)
Theorem A(3)(5)

Proposition 14.2(g)
$\left.\right\}$ $\rightarrow$ (10)
Theorem 15.7(a)

Theorem D

Corollary 15.3(b) $\rightarrow$ (1)
Lemma 12.17 $\rightarrow$ (2)

Theorem 14.4(b)
Theorem A(8) $\left.\right\}$ $\rightarrow$ (3)(4)
Corollary 15.9

Theorem E

Lemma 14.5(c) $\rightarrow$ (1)
Theorem 13.9 $\rightarrow$ (2)
Corollary 14.9 $\rightarrow$ (3)

Now we present the definition of the five types of maximal subgroups (essentially) as in **FT**. Some differences between them and between Theorem II below and Theorem 14.2 of **FT** will be discussed at the end of this section. We shall see that Type I corresponds to case $\mathscr{F}$, Type II to case $\mathscr{P}_2$, and that Types III, IV, and V constitute a refinement of case $\mathscr{P}_1$.

Let π^* be the set of all primes $p \in \pi(G)$ such that a Sylow p-subgroup P of G is cyclic or contains a subgroup A of order p such that $C_P(A) = A \times B$ with B cyclic. (By Lemma 10.13, the latter condition holds if P is not abelian and has a maximal elementary abelian subgroup of order p^2.) For the definition of Types I–V, let $H = M_F$. (As above, M_F denotes the maximal nilpotent normal Hall subgroup of M.) We say that M is of Type I if M enjoys the following properties:

(Ii) $1 \subset H \subset M$,
(Iii)   each complement E to H in M contains a normal abelian subgroup A such that $C_E(x) \subseteq A$ for all $x \in H^\#$,
(Iiii)   each complement E to H in M contains a subgroup E_0 of the same exponent as E such that HE_0 is a Frobenius group with Frobenius kernel H,
(Iiv)   every Sylow subgroup of M/H is abelian of rank at most 2,
(Iv) M satisfies at least one of the following conditions:
    (a) H is a TI-subset in G,
    (b) H is abelian of rank 2,
    (c) for every $p \in \pi(H)$, $p \in \pi^*$ and the exponent of M/H divides $p-1$; for some such prime p, we have $\mathcal{O}_{p'}(M)$ is cyclic.

Remark. In general a group M satisfying conditions (Ii)–(Iiii) with $H = M_F$ is called a group of *Frobenius type*, and H is called the *Frobenius kernel* of M.

We say that M is of Type II, III, IV, or V, if

(T1) M' is a Hall subgroup of M that contains H,
(T2)   a complement V of H in M' is nilpotent, and $N_M(V)$ has a cyclic subgroup W_1 of order $|M/M'|$,
(T3) H is not cyclic and $M'' \subseteq HC_M(H) = F(M) \subseteq M'$,
(T4) H contains a nonidentity cyclic subgroup W_2 such that $C_{M'}(x) = W_2$ for all $x \in W_1^\#$,
(T5)   if W_0 is any nonempty subset of the set $\widehat{W} = W_1 W_2 - (W_1 \cup W_2)$, then $N_G(W_0) = W_1 W_2$,
(T6)   if A_0 and A_1 are any two subgroups of prime order in V that are conjugate in G but not in M, then $C_H(A_0) = 1$ or $C_H(A_1) = 1$,
(T7)   for Type II–IV:
    (i) W_1 has prime order, and
    (ii) $F(M) = C_M(H)H$ is a TI-subset in G,

for Type II:
(IIiii) V is abelian of rank at most 2,
(IIiv) $V \neq 1$ and $N_G(V) \not\subseteq M$,
(IIv) $N_G(A) \subseteq M$ for every nonempty subset A of M' such that $C_H(A) \neq 1$,

for Type III:
(IIIiii) V is abelian and $N_G(V) \subseteq M$,

for Type IV:
(IViii) V is not abelian and $N_G(V) \subseteq M$,

for Type V:
(V) $V = 1$ and M satisfies at least one of the following conditions:
    (a) $M' = H$ is a TI-subset in G,
    (b) for some prime $p \in \pi(H) \cap \pi^*$, $O_{p'}(H)$ is cyclic and $|W_1| = |M/M'|$ divides $p - 1$,
    (c) for some prime $p \in \pi(H) \cap \pi^*$, $O_{p'}(H)$ is cyclic, and $|O_p(H)| = p^3$, and $|W_1|$ divides $p + 1$.

Remark. By using the existence of Hall subgroups in solvable groups and the nilpotence of solvable groups possessing fixed-point free automorphisms (Theorem 3.7), one can easily show that conditions (T1)–(T4) (for $H = M_F$) are equivalent to the definition of a *three step group* in **FT** (p. 780).

For the sake of clarity we prepare the proofs of the two main theorems (stated below) by the following proposition.

Proposition 16.1. (a) M is of Type I if and only if $M \in \mathcal{M}_{\mathscr{F}}$.
  (b) M is of Type II if and only if $M \in \mathcal{M}_{\mathscr{P}_2}$.
  (c) M is of Type III or IV if and only if $M \in \mathcal{M}_{\mathscr{P}_1}$ and $M_F \neq M_\sigma$.
  (d) M is of Type V if and only if $M \in \mathcal{M}_{\mathscr{P}_1}$ and $M_F = M_\sigma$.
  (e) $M' = U M_\sigma$ if and only if M is not of Type I.
  (f) $M_F = M_\sigma$ if and only if M is of Type I, II, or V.

Proof. Let $H = M_F$. By Theorem A(6),

$$1 \subset H \subseteq M_\sigma \subseteq M' \subset M.$$

Suppose $M \in \mathcal{M}_{\mathscr{F}}$, i.e., $K = 1$ and $U \neq 1$. By Theorem A(8), $H = M_\sigma$. Thus U complements H in M. Therefore Theorem B(1), (2), and (3) show that M satisfies conditions (Ii)–(Iiv), and Theorem 15.7(c) yields condition (Iv). Thus M is of Type I.

Conversely, if M is of Type I, but not of type $\mathscr{F}$, then $K \neq 1$ and $1 \subset K^* = C_H(K)$ by Theorem C(2). On the other hand, since we know that $K \cap H \subseteq K \cap M_\sigma = 1$ and that K is a nonidentity cyclic Hall subgroup of M, condition (Iiii) implies that $C_H(K) = 1$. This contradiction completes the proof of (a).

Now we may assume that $M \in \mathcal{M}_{\mathscr{P}}$, i.e., $K \neq 1$. By Theorem C(3), $M' = UM_\sigma$. Thus M' is a Hall subgroup of M that contains H and complements the cyclic subgroup K. Define

$$W_1 = K \text{ and } W_2 = K^* \quad (= C_{M_\sigma}(K))$$

and let V be a K-invariant complement of H in M'. We may let $V = U$ if $M_\sigma = H$.

By Theorem A(6) and (7), V is nilpotent and

$$M'' \subseteq F(M) = C_M(H)H \subseteq M'.$$

By Theorem C(2), $W_2 = K^*$ is cyclic, $1 \subset K^* \subseteq H$, and M_F is not cyclic. Thus M satisfies conditions (T1)–(T3) in the definition of Types II–V, and (T4) and (T5) follow directly from Theorem A(5) and Theorem C(8), respectively.

Suppose A_0 and $A_1 = A_0{}^g$ $(g \in G)$ are as in (T6), with $C_H(A_i) \neq 1$ $(i = 1, 2)$. If $H = M_\sigma$, then $V = U$ and Theorem B(4) shows that $\{M\} = \mathcal{M}(C_G(A_1)) = \mathcal{M}(C_G(A_0))^g = \{M\}^g$, and hence $g \in N_G(M) = M$, a contradiction. Therefore $H \neq M_\sigma$. By Theorem A(8),

$$U = 1 \text{ and } A_i \subseteq V \subseteq M' = UM_\sigma = M_\sigma.$$

Now Theorem D(1) shows that A_0 and A_1 are conjugate in M, again a contradiction.

This completes the proof of conditions (T1)–(T6).

Assume $K \neq 1$ and $U \neq 1$, i.e., $M \in \mathcal{M}_{\mathscr{P}_2}$. Then $V = U$ because $M_\sigma = H$ by Theorem A(8), and conditions (i)–(IIv) in (T7) follow directly from Theorem C(1) and (10) and Theorem B(1) and (4). Thus M is of Type II.

Assume next that $K \neq 1$ and $U = 1$, i.e., $M \in \mathcal{M}_{\mathscr{P}_1}$. Then $V \subseteq M' = M_\sigma$. Suppose $V \neq 1$, i.e., $H \neq M_\sigma$. Then, again by Theorem A(8), $W_1 = K$ has prime order and $F(M)$ is a TI-subset of G. Furthermore, there exists a prime $p \in \pi(V) \cap \sigma(M)$ and V contains a Sylow p-subgroup P of M because V is a Hall subgroup of M. By the Frattini argument and the definition of $\sigma(M)$,

$$N_G(V) \subseteq VN_G(P) \subseteq M.$$

This proves that M is of Type III or IV if $M \in \mathcal{M}_{\mathscr{P}_1}$ and $H \neq M_\sigma$.

Finally, if $K \neq 1$ and $U = V = 1$, then Theorem 15.7(c) yields condition (T7)(V), and M is of Type V.

In order to complete the proof of (b), (c), and (d), note that if M is of Type II, III, IV, or V, then $\pi(W_1) \subseteq \kappa(M)$ because $W_1 \cap M_\sigma \subseteq W_1 \cap M' = 1$, W_1 is cyclic, and $1 \subset C_H(W_1) \subseteq C_{M_\sigma}(W_1)$. Thus $\kappa(M)$ is not empty, and this means that $M \in \mathcal{M}_{\mathscr{P}}$. Now (b), (c), and (d) follow from the implications proved above and from the obvious fact that M cannot belong to two of the Types II–V.

Since $M' = UM_\sigma$ if $M \in \mathcal{M}_\mathscr{P}$, and $UM_\sigma = M \supset M'$ if $M \in \mathcal{M}_\mathscr{F}$, (e) holds.

If $H \neq M_\sigma$, then $M \in M_{\mathscr{P}_1}$ by Theorem 15.2(a), and thus (c) and (d) yield (f). $\square$

Now this proposition together with Theorem C(4), (6), and (7) yield the following first main theorem of **FT**, except the first assertion, which follows directly from Corollaries 15.4 and 15.3(b).

Theorem I. Two elements of a nilpotent Hall subgroup H of G are conjugate in G if and only if they are conjugate in $N_G(H)$.

Either every maximal subgroup of G is of Type I or all of the following conditions are true.

(a) G contains a cyclic subgroup $W = W_1 \times W_2$ with the property that $N_G(W_0) = W$ for every nonempty subset W_0 of $W - W_1 - W_2$. Also, $W_i \neq 1$ $(i = 1, 2)$.

(b) G contains maximal subgroups S and T not of Type I such that $S = W_1 S'$, $T = W_2 T'$, $S' \cap W_1 = 1$, $T' \cap W_2 = 1$, and $S \cap T = W$.

(c) Every maximal subgroup of G is either conjugate to S or T or is of Type I.

(d) S or T is of Type II.

(e) Both S and T are of Type II, III, IV or V (they are not necessarily of the same type).

The following characterization of the sets $A(M)$ and $A_0(M)$, which uses the notation of the definition of the Types I–V, is a direct consequence of Proposition 16.1 and Theorem C(9). Actually, it is the original definition of $A(M)$ and $A_0(M)$ on p. 847 of **FT**, where they are called $\widehat{M}$ and $\widehat{M}_1$, respectively. Recall that $H = M_F$.

$$A(M) = A_0(M) = \bigcup_{x \in H^\#} C_M(x) \qquad \text{if } M \text{ is of Type I,}$$

$$A(M) = \bigcup_{x \in H^\#} C_{M'}(x) \qquad \text{if } M \text{ is of Type II,}$$

$$A(M) = M' \qquad \text{if } M \text{ is of Type III, IV, or V,}$$

$$A_0(M) = A(M) \bigcup \mathscr{C}_M(\widehat{W}) \qquad \text{if } M \text{ is of Type II, III, IV, or V.}$$

The next theorem is concerned with the embedding of $A(M)$ and $A_0(M)$ in G.

Theorem II. For an arbitrary maximal subgroup M of G, let $X = A(M)$ or $X = A_0(M)$, and let $D = \{\, x \in X^\# \mid C_G(x) \nsubseteq M \,\}$.

Then $D \subseteq A(M)$, $|\mathscr{M}(C_G(x))| = 1$ for all $x \in D$, and the following conditions are satisfied:

(Ti) Whenever two elements of X are conjugate in G, they are conjugate in M.

(Tii) If D is not empty, then there are maximal subgroups $M_1, \ldots, M_n$ of G of Type I or II such that, with $H_i = M_{iF}$ (and therefore $H_i \subseteq M_i'$).

   (a) $(|H_i|, |H_j|) = 1$ for $i \neq j$,
   (b) $M_i = H_i(M \cap M_i)$ and $M \cap H_i = 1$,
   (c) $(|H_i|, |C_M(x)|) = 1$ for all $x \in X^\#$,
   (d) $A_0(M_i) - H_i$ is a nonempty TI-subset in G with normalizer M_i, and
   (e) if $x \in D$, then there is a conjugate y of x in D and an index i such that $C_G(y) = C_{H_i}(y)C_M(y) \subseteq M_i$.

(Tiii) If some M_i in (Tii) has Type II, then M is a Frobenius group (thus of Type I) with cyclic Frobenius complement, and M_F is not a TI-subset in G.

Proof. The set $A_0(M)$ is the disjoint union of the sets M_σ, $A(M) - M_\sigma$, and $A_0(M) - A(M)$. An element of any of these sets is not conjugate to an element of one of the other two sets because the orders of these two elements are distinct. Moreover, by Theorem B(5) and Theorem C(9), the latter two sets are TI-subsets of G (with normalizer M if not empty). Therefore

$$D = \{\, x \in M_\sigma^\# \mid C_G(x) \nsubseteq M \,\} \subseteq M_\sigma,$$

and (Ti) reduces to the statement that any two elements of M_σ conjugate in G are already conjugate in M, which is true by Theorem D(1).

So assume that D is not empty. If $x \in D$, then $C_G(x)$ lies in a unique maximal subgroup $N(x)$ of G, and $N(x)$ is of Type I or Type II by Theorem D(4). Let $\mathscr{A}$ be the collection of all such subgroups $N(x)$ and let $\{\, M_1, \ldots, M_n \,\}$ be a subset of $\mathscr{A}$ such that each $N \in \mathscr{A}$ is conjugate in G to exactly one M_i. Now (Tiii) follows from Theorem D(4). So we must prove (Tii).

Take some M_i. By Theorem D(4), $H_i = M_{iF} = M_{i\sigma}$ and (b) holds. By Theorem E(2), the sets $\sigma(M_i)$ are pairwise disjoint, which implies (a) because $\sigma(M_i) = \pi(M_{i\sigma})$. Also, by Theorem D(4), $A(M_i) - M_{i\sigma}$ is not empty and, by Theorem B(5), this is a TI-subset of G (with normalizer M_i). If M_i is of Type I, this gives us (d). Otherwise, by Theorem C(5) and a short argument, $A_0(M_i) - A(M_i)$ and $A_0(M_i)$ are TI-subsets of G with normalizer M_i.

If $x \in D$, then there exists an element $g \in G$ and an index i such that $N(x)^g = M_i$. By Lemma 14.13(b), we can choose g in M. Then $y = x^g \in D$, $M_i = N(y)$, and $C_G(y) = C_{H_i}(y)C_{M \cap M_i}(y)$ by Theorem D(3) and (4).

It remains to prove (c). Suppose $x \in X^\#$ and $(|H_i|, |C_M(x)|) \neq 1$. Then $\sigma(M_i) \cap \pi(M)$ is not empty. By Lemma 14.13(a), M is a Frobenius group with kernel M_σ. This means that $A_0(M) = M_\sigma$. Therefore $x \in M_\sigma$ and $C_M(x) \subseteq A_0(M) = M_\sigma$. Consequently $\sigma(M_i) \cap \sigma(M)$ is not empty. By Theorem E(2), this implies that M_i is conjugate to M in G, a contradiction (since $\tau_2(M)$ is empty, while $\tau_2(N) \supseteq \pi(\langle x \rangle)$ by Theorem 14.4(c)). $\square$

Remark. A subset X of G satisfying conditions (Ti)–(Tiii) of Theorem II is here called a *tamely imbedded subset* of G and the subgroups H_i are said to form a *system of supporting subgroups* for X. The realization that the concept of a TI-subset could be fruitfully extended to the more general concept of a tamely imbedded subset was one of the major achievements of Feit and Thompson. As the proof shows, the essence of Theorem II is that M_σ is tamely imbedded in G. Note that if D is empty, then X is a TI-subset in G.

Theorem II above corresponds to Theorem 14.2 of **FT** and differs from it in two significant respects. First, Theorem II includes (Tiii) and the condition that each M_i in (Tii) be of Type I or II. These conditions were kindly communicated to us by Feit and Thompson and by David Sibley, respectively. Second, condition (Tii)(d) replaces a stronger condition (Definition 9.1(ii)(e), pp. 803–804) in the definition of a tamely imbedded subset in **FT**, which states that

$$(16.1) \qquad \text{for } \widehat{M_i} = \left\{ \bigcup_{x \in H_i \#} C_{M_i}(x) \right\} - H_i,$$

$\widehat{M_i}$ is a nonempty TI-subset in G with normalizer M_i.

This amounts to M_i being of Type I, a fact we cannot prove. However, Feit and Thompson have informed us that condition (Tiii) suffices for Chapter V (specifically Section 33) in **FT**.

As to the definition of Types I–V, conditions (Iv)(c) and (T7)(V)(b) and (c) are a little more explicit than the corresponding conditions in **FT**. Second, condition (T7)(IIv) is also included in the definition of Types III and IV in **FT**, but is apparently not used later. Furthermore, there is also a slight difference between our definition of Type I (and of Frobenius type) and that in **FT**. There we find the stability groups $I_E(\theta)$ of the nontrivial irreducible characters θ of H in place of the centralizers $C_E(x)$ for $x \in H^\#$. However it is well known and an easy exercise in elementary character theory that if H is a normal Hall subgroup of a finite group M, then an element $x \in M$ fixes some nontrivial irreducible character of H if and only if $C_H(x) \neq 1$.

APPENDIX A

Prerequisites and p-Stability

Among the main tools for shortening the first half of the proof of **FT** are Theorems 6.1 and 6.2, which are obtained by use of the concept of *p-stability*. In Section 6 these are obtained from theorems in **G**, which have shorter proofs if one restricts to groups of odd order and uses a different characteristic subgroup in place of $J(S)$. In this appendix and Appendix B we outline these shorter proofs. Although we use some results from Chapters 1–6 of **G**, this makes it unnecessary to use some other results from **G**, as described below.

This appendix is devoted mainly to proving Theorem 6.1 and a special case of Theorem 6.5.3 of **G** that will be applied in Appendix B. For those who wish to read both this appendix and Appendix B, the prerequisites for this book may be reduced and handled as follows. One first reads Chapters 1–6 and Section 7.3 of **G**, except for Theorems 2.8.3 and 2.8.4 (pp. 42–55) and Sections 3.8 and 6.5. Next one reads Sections 1 and 2 in Chapter I of this book, followed by this appendix and Appendix B (including parts of Sections 3.8 and 6.5 of **G** mentioned later in this appendix). In particular, one does not need to read Chapter 8 and most of Chapter 7 of **G**.

Additional prerequisites for the proof of the CN-theorem are described in Appendix D.

To begin, we refer the reader to pages 39–40 of **G**, which introduce the groups $\mathbf{GL}(2, q)$ and several related families of groups. This is followed in **G** by an important theorem of Dickson (Theorem 2.8.4) with a rather long, complicated proof. Fortunately, we require only an easy corollary of Dickson's Theorem, which we prove directly.

Theorem A.1. Suppose V is a 2-dimensional vector space over a field **F** of odd characteristic p and G is a finite, irreducible group of linear transformations of V over **F** such that $|G|$ is odd.

Then p does not divide $|G|$.

Proof. By Theorem 2.6 of this book, G has an abelian Sylow p-subgroup N that contains G'. Then $N \lhd G$. By **G**, Theorem 3.1.3, p. 62, $N = 1$. $\quad\square$

Now we move to Section 3.8 of **G**. In place of Theorem 3.8.1 there, we use the following result, which has the same hypothesis, but a weaker conclusion.

Theorem A.2. Let p be an odd prime. Let G be a group of linear transformations acting faithfully and irreducibly on a vector space V over an algebraic closure of $\mathbf{F}_p$. Assume that G is generated by two p-elements which have a quadratic minimal polynomial on V.

Then G has even order.

Proof. Follow the proof of Theorem 3.8.1 of **G** up to the point, on page 105, where V is shown to have dimension 2. Since G is generated by two p-elements, Theorem A.1 shows that $|G|$ is even. $\quad\square$

By continuing to the top of page 109 of **G**, but using Theorem A.2 in place of Theorem 3.8.1 of **G**, we obtain the following substitute for Theorem 3.8.3 of **G**. (Incidentally, some very short proofs of Theorem 3.8.2 of **G** are given in [10, pp. 4–5].)

Theorem A.3. Let p be an odd prime and G be a group with no nontrivial p-subgroups. If G is not p-stable, then G has even order.

We now move to Section 6.5 of **G**. By using Theorem A.3 instead of Theorem 3.8.4(e), we obtain special cases of Theorems 6.5.1–6.5.3, which we state as follows.

Theorem A.4. Let p be an odd prime and G be a solvable group of odd order. Let P be a p-subgroup of G.

    (a) If $\mathcal{O}_p(G) = 1$, then G is p-stable.

    (b) If P is a Sylow p-subgroup of G, then every normal abelian subgroup of P is contained in $\mathcal{O}_{p',p}(G)$.

    (c) Suppose $\mathcal{O}_{p'}(G)P \lhd G$ and A is a p-subgroup of $N_G(P)$ for which $[P, A, A] = 1$. Then $AC_G(P)/C_G(P) \subseteq \mathcal{O}_p(N_G(P)/C_G(P))$.

Note that Theorem A.4(b) is just Theorem 6.1. We now apply (c).

Theorem A.5. Suppose p is an odd prime, G is a solvable group of odd order, P is a normal p-subgroup of G, and X is a subgroup of G that is generated by abelian p-groups, each of which is normalized by P. Then

    (1) $XC_G(P)/C_G(P) \subseteq \mathcal{O}_p(G/C_G(P))$, and

    (2) if $\mathcal{O}_{p'}(G) = 1$ and $C_{\mathcal{O}_p(G)}(P) \subseteq P$, then $X \subseteq \mathcal{O}_p(G)$.

Proof. Note that $G = N_G(P)$. Let $C = C_G(P)$. For each abelian p-subgroup A of G normalized by P we have

$$AC/C \subseteq \mathcal{O}_p(G/C),$$

by Theorem A.4(c) (since $[P, A, A] \subseteq [A, A] = 1$). Therefore

$$XC/C \subseteq \mathcal{O}_p(G/C),$$

which is (1).

Next let $Q = \mathcal{O}_p(G)$ and assume that $\mathcal{O}_{p'}(G) = 1$ and $C \cap Q \subseteq P$. Then $P \subseteq Q$ and $Q = \mathcal{O}_{p',p}(G)$. By Proposition 1.15(b), $C_G(Q) \subseteq \mathcal{O}_{p',p}(G) = Q$. For every p'-element u in C,

$$C_Q(C_Q(u)) \subseteq C_Q(P) = C \cap Q \subseteq P \subseteq C_Q(u),$$

so that u centralizes Q by Proposition 1.8. As $C_G(Q) \subseteq Q$, it follows that $u = 1$. Thus C is a p-group. Moreover, $C \subseteq \mathcal{O}_p(G) = Q$ because $C \lhd G$.

Since $C \subseteq \mathcal{O}_p(G)$ we have $\mathcal{O}_p(G/C) = \mathcal{O}_p(G)/C$. By (1),

$$XC/C \subseteq \mathcal{O}_p(G/C) = \mathcal{O}_p(G)/C,$$

which yields (2). $\square$

APPENDIX B

The Puig Subgroup

In this appendix we will define an important characteristic subgroup of a finite group G and derive some of its remarkable properties (some of which are similar to the properties of the Thompson J-subgroup described in **G**, Chapter 8).

Our main goal is a result of L. Puig (analogous to **G**, Theorem 8.2.11, p. 279 about $Z(J(S))$), which we present using a short, unpublished proof of I. M. Isaacs.

Throughout this appendix we will assume that G is a finite group and p is a prime. We will use the fact that solvable groups of odd order have the nice properties described in the following theorem from Appendix A.

Theorem A.5. Suppose p is an odd prime, G is a solvable group of odd order, P is a normal p-subgroup of G, and X is a subgroup of G that is generated by abelian p-groups each of which is normalized by P. Then

(1) $XC_G(P)/C_G(P) \subseteq \mathcal{O}_p(G/C_G(P))$, and
(2) if $\mathcal{O}_{p'}(G) = 1$ and $C_{\mathcal{O}_p(G)}(P) \subseteq P$, then $X \subseteq \mathcal{O}_p(G)$.

Remark. By using the original theorems in **G** rather than the theorems in Appendix A, one can generalize this theorem by substituting for the odd order condition the requirement that G have abelian or dihedral Sylow 2-subgroups or more generally that $\mathbf{SL}(2,p)$ not be involved in G.

Notation. For subgroups X and Y of any group G, we write $X \rightarrow Y$ if Y is generated by abelian subgroups of G that are all normalized by X. Clearly, given a subgroup X of G, there is a unique largest subgroup Y of G such that $X \rightarrow Y$. We will denote this subgroup by $L_G(X)$.

We define recursively the following sequence of subgroups of an arbitrary group G:

$$L_n(G) = \begin{cases} 1, & \text{if } n = 0 \\ L_G(L_{n-1}(G)), & \text{if } n > 0. \end{cases}$$

Thus we have

$$L_0(G) \to L_1(G) \to L_2(G) \to \cdots \to L_n(G) \to L_{n+1}(G) \to \ldots.$$

Finally, define

$$L(G) = \bigcap_{n \geq 0} L_{2n+1}(G) \quad \text{and} \quad L_*(G) = \bigcup_{n \geq 0} L_{2n}(G).$$

Lemma B.1. Let G be any finite group. Then the subgroups $L_i(G)$ have the following properties.

(a) If $X \subseteq Y \subseteq G$, then $L_G(X) \supseteq L_G(Y)$.

(b)

$$1 = L_0(G) \subseteq L_2(G) \subseteq L_4(G) \subseteq \cdots$$
$$\cdots \subseteq L_5(G) \subseteq L_3(G) \subseteq L_1(G) = G$$

(c) For some $k \geq 0$,

$$L_{2n}(G) = L_*(G) \text{ for all } n \geq k, \text{ and}$$
$$L_{2n+1}(G) = L(G) \text{ for all } n \geq k.$$

(d) The set $L_*(G)$ is a subgroup of G and is contained in $L(G)$.

(e) Every abelian normal subgroup of G is contained in $L_i(G)$ for all positive integers i.

(f) If G is a p-group, then, for all $i > 0$,

$$L_i(G) \supseteq C_G(L_i(G)) \supseteq Z(G);$$
$$L_*(G) \supseteq C_G(L_*(G)) \supseteq Z(G); \text{ and}$$
$$L(G) \supseteq C_G(L(G)) \supseteq Z(G).$$

In particular, if $G \neq 1$, then $L_*(G) \neq 1$ and $L(G) \neq 1$.

(g) $L_G(L_*(G)) = L(G)$ and $L_G(L(G)) = L_*(G)$.

Proof. (a) Suppose that $X \subseteq Y$. Then clearly $X \to L_G(Y)$ and hence $L_G(X) \supseteq L_G(Y)$.

(b) By definition, $L_0(G) = 1$. Since every element of G lies in a cyclic group, $L_1(G) = G$. Trivially,

$$L_0(G) \subseteq L_2(G) \subseteq L_1(G).$$

By (a),

$$L_2(G) \subseteq L_3(G) \subseteq L_1(G).$$

Thus an easy induction argument yields, for all integers $n \geq 0$,

$$L_{2n}(G) \subseteq L_{2n+2}(G) \subseteq L_{2n+1}(G) \text{ and}$$
$$L_{2n+2}(G) \subseteq L_{2n+3}(G) \subseteq L_{2n+1}(G).$$

Combining these results yields (b).

(c), (d) Since G is a finite group, the increasing sequence of subgroups $\{L_{2i}(G)\}$ must stabilize at some integer k_1. For this k_1 we have

$$L_{2n}(G) = L_*(G) \text{ for all } n \geq k_1.$$

Similarly, the decreasing sequence $\{L_{2i+1}(G)\}$ must stabilize at some integer k_2, and

$$L_{2n+1}(G) = L(G) \text{ for all } n \geq k_2.$$

Choosing $k = \max\{k_1, k_2\}$ clearly gives us (c) and (d).

(e) In fact, if A is an abelian normal subgroup of G and if H is any subgroup of G, then $H \to A$ and hence $A \subseteq L_G(H)$. Since $L_i(G) = L_G(L_{i-1}(G))$ for all positive integers i, we have (e).

(f) In view of (c), we need only prove the first statement of (f). Clearly $Z(G) \subseteq C_G(L_i(G))$. Take any $i \geq 1$ and let M be a normal subgroup of G maximal subject to being abelian. Then, by (e), $M \subseteq L_i(G)$. Furthermore, since G is a p-group, by **G**, Theorem 5.3.12, p. 185, $C_G(M) \subseteq M$. Thus

$$C_G(L_i(G)) \subseteq C_G(M) \subseteq M \subseteq L_i(G).$$

This yields (f).

(g) This follows immediately from (c). □

Lemma B.2. Suppose H is a subgroup of G that contains $L(G)$. Then $L(G) = L(H)$.

Proof.

Step 1. (a) $L_{2n+1}(H) \subseteq L_{2n+1}(G)$ for all $n \geq 0$;
  (b) $L(H) \subseteq L(G)$.

Proof. We will use induction on n to prove (a). For $n = 0$ we have

$$H = L_1(H) \subseteq L_1(G) = G.$$

Assume now that $n > 0$ and $L_{2n-1}(H) \subseteq L_{2n-1}(G)$. Then

$$L_{2n-1}(H) \to L_G(L_{2n-1}(G)) = L_{2n}(G) \subseteq L(G) \subseteq H.$$

Thus

$$L_{2n}(G) \subseteq L_H(L_{2n-1}(H)) = L_{2n}(H).$$

Therefore

$$L_{2n}(G) \to L_{2n+1}(H).$$

Consequently

$$L_{2n+1}(H) \subseteq L_{2n+1}(G).$$

This yields (a). Now Lemma B.1(c) gives us (b). □

Step 2. (a) $L(G) \subseteq L_{2n+1}(H)$ for all $n \geq 0$;
  (b) $L(G) \subseteq L(H)$.

Proof. Again we use induction for (a). For $n = 0$ we know $L_1(H) = H$, so (a) is true by hypothesis. Now suppose that $n > 0$ and $L(G) \subseteq L_{2n-1}(H)$. Then

$$L(G) \to L_{2n}(H).$$

Hence

$$L_{2n}(H) \subseteq L_G(L(G)) = L_*(G).$$

Furthermore,

$$L_{2n}(H) \to L_G(L_*(G)) = L(G) \subseteq H.$$

Thus

$$L(G) \subseteq L_{2n+1}(H).$$

Again, (b) follows from Lemma B.1(c). $\square$

By Step 1 and Step 2 we can now conclude that $L(G) = L(H)$ as desired. $\square$

Lemma B.3. Assume p is odd, G is solvable of odd order, and $\mathcal{O}_{p'}(G) = 1$. Suppose that S is a Sylow p-subgroup of G and $T = \mathcal{O}_p(G)$. Then

$$L_*(S) \subseteq L_*(T) \subseteq L(T) \subseteq L(S).$$

Proof. First we show by induction on n that for all $n \geq 0$,

(B.1) $$L_{2n}(S) \subseteq L_{2n}(T) \subseteq L_{2n+1}(T) \subseteq L_{2n+1}(S).$$

For $n = 0$ the statement reduces to

$$1 \subseteq 1 \subseteq T \subseteq S,$$

which is trivial.

Assume (B.1) holds for some n. Since $L_{2n+1}(S) \to L_{2n+2}(S)$, we get

(B.2) $$L_{2n+1}(T) \to L_{2n+2}(S).$$

Now $L_{2n+1}(T)$ is a normal p-subgroup of G and, by Lemma B.1(f),

$$L_{2n+1}(T) \supseteq C_T(L_{2n+1}(T)).$$

Thus, by (B.2) and Theorem A.5,

$$L_{2n+2}(S) \subseteq T.$$

Hence, by (B.2),

(B.3) $$L_{2n+2}(S) \subseteq L_T(L_{2n+1}(T)) = L_{2n+2}(T).$$

Consequently, by Lemma B.1(a),

(B.4) $$L_{2n+3}(T) = L_T(L_{2n+2}(T)) \subseteq L_T(L_{2n+2}(S))$$
$$\subseteq L_S(L_{2n+2}(S)) = L_{2n+3}(S).$$

By Lemma B.1(b),

$$L_{2n+2}(T) \subseteq L_{2n+3}(T)$$

so, by (B.3) and (B.4),

$$L_{2n+2}(S) \subseteq L_{2n+2}(T) \subseteq L_{2n+3}(T) \subseteq L_{2n+3}(S).$$

Thus we have (B.1) for $n+1$ in place of n. Now, by Lemma B.1(c), we have the conclusion of the lemma. $\square$

Theorem B.4 (L. Puig, 1976). Assume that p is odd, G is solvable of odd order, and S is a Sylow p-subgroup of G. Then

(a) $G = \mathcal{O}_{p'}(G)N_G(Z(L(S)))$;
(b) if $\mathcal{O}_{p'}(G) = 1$, then $Z(L(S)) \lhd G$.

Remark. Note that Theorem B.4(a) serves as a substitute for Theorem 6.2 of this work.

Proof. Let $L = L(S)$, $T = \mathcal{O}_p(G)$, and $Y = Z(L(T))$.

Step 1. Part (b) implies part (a).

Proof. Let $\overline{G} = G/\mathcal{O}_{p'}(G)$ and $\overline{S} = S\mathcal{O}_{p'}(G)/\mathcal{O}_{p'}(G)$. Then, by **G**, Theorem 6.3.1(iv), p. 227,

$$\mathcal{O}_{p'}(\overline{G}) = \mathcal{O}_{p'}(G/\mathcal{O}_{p'}(G)) = 1.$$

Let $Z = Z(L(S))$ and $\overline{Z} = Z\mathcal{O}_{p'}(G)/\mathcal{O}_{p'}(G)$. Then $\overline{Z} = Z(L(\overline{S}))$ because $\overline{S} \cong S$. Thus, assuming (b), we know that $\overline{Z} \lhd \overline{G}$ and hence $Z\mathcal{O}_{p'}(G) \lhd G$. Clearly Z is a Sylow p-subgroup of $Z\mathcal{O}_{p'}(G)$. Consequently, by the Frattini argument, we have

$$G = Z\mathcal{O}_{p'}(G)N_G(Z) = \mathcal{O}_{p'}(G)ZN_G(Z) = \mathcal{O}_{p'}(G)N_G(Z). \quad \square$$

We will henceforth assume that $\mathcal{O}_{p'}(G) = 1$.

Step 2. $Z(L(S)) \subseteq Y$.

Proof. By Lemma B.3,

(B.5) $L_*(S) \subseteq L_*(T) \subseteq L(T) \subseteq L(S)$.

Thus, by Lemma B.1(f),

$$L(T) \supseteq L_*(S) \supseteq C_S(L_*(S)) \supseteq Z(L(S)).$$

Consequently, by (B.5),

$$Z(L(S)) \subseteq Z(L(T)) = Y. \quad \square$$

Step 3. Let $C \subseteq G$ be taken such that $C/C_G(Y) = \mathcal{O}_p(G/C_G(Y))$. Then $L(S) \lhd N_G(C \cap S)$.

Proof. Since Y char T, we know that $C_G(Y) \lhd G$ and $C \lhd G$. As Y is abelian and $Y \lhd S$, Lemma B.1 yields

$$Y \subseteq L_*(S).$$

Now $L(S)$, $L_*(S)$ char S and $L(T)$, $L_*(T) \lhd G$. Since Lemma B.1(g) tells us that $L = L_G(L_*(S))$, we have

$$L_*(S) \to L$$

and hence

$$Y \to L.$$

By Theorem A.5,

$$L C_G(Y)/C_G(Y) \subseteq \mathcal{O}_p(G/C_G(Y)) = C/C_G(Y).$$

Therefore $L \subseteq C$ and $L \subseteq C \cap S$. Finally, by Lemma B.2,

$$L = L(C \cap S) \lhd N_G(C \cap S). \quad \square$$

Step 4. Conclusion.

Proof. Take C as in the previous step. Then $C \cap S$ is a Sylow p-subgroup of C and $(C \cap S)C_G(Y)/C_G(Y)$ is a Sylow p-subgroup of $C/C_G(Y)$. But $C/C_G(Y)$ is a p-group, so $(C \cap S)C_G(Y)/C_G(Y) = C/C_G(Y)$ and we can conclude that

$$C = (C \cap S)C_G(Y) = C_G(Y)(C \cap S).$$

By the Frattini argument,

$$
\begin{aligned}
\text{(B.6)} \quad G = C N_G(C \cap S) &= C_G(Y)(C \cap S)N_G(C \cap S) \\
&= C_G(Y)N_G(C \cap S).
\end{aligned}
$$

Since, by Step 2, $Z(L) \subseteq Y$, we have $Z(L) \subseteq Z(C_G(Y))$. By Step 3, $L \lhd N_G(C \cap S)$, and consequently

$$Z(L) \lhd N_G(C \cap S).$$

Now, by (B.6), we can conclude that $Z(L) \lhd G$ as desired. $\quad \square$

APPENDIX C

The Final Contradiction

In the original paper of Feit and Thompson, the minimal counterexample is studied by means of local analysis (as in the present work), then by character theory, and finally by a relatively short (17 page) argument using generators and relations which produces a contradiction. This last argument was substantially simplified by Thomas Peterfalvi in [22] (using also some previous reductions of R. Howlett, L. G. Kovacs, M. F. Newman, and the second author). In this appendix we present, with slight changes, an account of Peterfalvi's work written by Walter Carlip and Wayne W. Wheeler at the University of Chicago for the Junior Group Theory Seminar [2].

1. The Main Theorem

Theorem C. Let p and q be two primes satisfying the following condition:

(A) $\left(\dfrac{p^q - 1}{p - 1}, p - 1\right) = 1.$

Let P be the additive group of the field $\mathbf{F}_{p^q}$ and U the subgroup of the multiplicative group $(\mathbf{F}_{p^q})^*$ consisting of the elements of norm one over $\mathbf{F}_p$. The subgroup U acts on P by multiplication and we can form the semidirect product $H = PU$. Let P_0 be the image in P of the additive group of $\mathbf{F}_p$. Furthermore, suppose that there is a group G such that hypothesis (B) below holds.

(B) There is a monomorphism $\sigma \colon H \to G$, a finite abelian p'-subgroup Q of G, and an element $y \in Q$ such that $\sigma(P_0)$ normalizes Q and $\sigma(P_0)^y$ normalizes U.

Then $p \leq q$.

2. Preliminary Remarks and Results

(I) Condition (A) for primes p and q is equivalent to the condition that q not divide $p - 1$. To see this, let $r = p - 1$. Then

$$p^q - 1 \equiv (r + 1)^q - 1 \equiv (r^q + qr^{q-1} + \cdots + qr + 1) - 1 \equiv qr \pmod{r^2}$$

and hence $(p^q - 1)/r \equiv q \pmod{r}$.

(II) (T. Peterfalvi) Take any prime q. A short argument shows that the hypothesis of the theorem is satisfied by taking $p = 2$, $G = \mathbf{SL}(2, 2^q)$, and

$$\sigma(P) = \left\{ \begin{pmatrix} 1 & a \\ 0 & 1 \end{pmatrix} \middle| a \in \mathbf{F}_{p^q} \right\}, \qquad \sigma(P_0) = \left\langle \begin{pmatrix} 1 & 1 \\ 0 & 1 \end{pmatrix} \right\rangle$$

$$\sigma(U) = \left\{ \begin{pmatrix} a & 0 \\ 0 & a^{-1} \end{pmatrix} \middle| a \in (\mathbf{F}_{p^q})^* \right\},$$

$$y = \begin{pmatrix} 0 & 1 \\ 1 & 1 \end{pmatrix}, \qquad \text{and} \qquad Q = \langle y \rangle.$$

(III) If G is a minimal counterexample to the Feit-Thompson Theorem, then there are primes p and q such that p and q satisfy (A), G satisfies (B), and $p > q$. (This can be derived from Theorem 27.1 and Lemma 38.1, pp. 943 and 1101 of **FT**.)

(IV) In [12], S. P. Norton and the second author have extended Theorem C to show that $p \leq 3$. Example (II) above shows that p may be equal to 2. It is not yet known whether p may be equal to 3.

(V) By (A), we can assume that p and q are odd.

(VI) We will identify H with its image in G and write the operation in P multiplicatively.

(VII) The following facts from Galois theory can be found in Theorem 38, p. 46 and Theorem 33, p. 42 of [20]. The second of these is Hilbert's famous Satz 90.

$$\mathrm{Gal}(\mathbf{F}_{p^q}/\mathbf{F}_p) = \langle \alpha \rangle, \text{ where } x^\alpha = x^p \text{ and}$$

$$U = \left\{ \frac{x}{x^\alpha} \middle| x \in \mathbf{F}_{p^q}^* \right\} \text{ so } U \text{ is cyclic of order } \frac{p^q - 1}{p - 1}.$$

(VIII) $\mathbf{F}_{p^q}^* = \mathbf{F}_p^* \times U$ by (A).

(IX) H is a Frobenius group with kernel P and complement U.

(X) $Q = C_Q(P_0) \times [Q, P_0]$ by an extension of **G**, Theorem 5.2.3, p. 177.

(XI) By (X) we may assume that $y \in [Q, P_0]$.

Notation. We will denote the norm of an element $a \in \mathbf{F}_{p^q}$ over $\mathbf{F}_p$ by $N(a)$. Also we will denote by E the following set:

$$E = \{ a \in \mathbf{F}_{p^q} \mid N(a) = N(2 - a) = 1 \}.$$

3. Proof of the Main Theorem

Here we begin the proof of Theorem C. Assume the hypotheses of Theorem C and, as in (V), assume that p and q are odd.

Lemma C.1. If $E = E^{-1}$ and $|E| \geq 2$ then $p \leq q$.

Proof. Let $a \in E^{\#}$. Then $2 - a \in E$ and $\tau(a) = 1/(2-a) \in E$ since $E = E^{-1}$. By induction, $\tau^k(a) \in E$ and $\tau^k(a) = [k-(k-1)a]/[(k+1)-ka]$ for every natural number k. Applying this formula for $\tau^k(a)$ we have

$$1 = \mathbf{N}\left(\frac{1}{\tau(a)\tau^2(a)\cdots\tau^k(a)} \right) = \mathbf{N}((k+1) - ka) = \mathbf{N}((1-a)k + 1)$$

for all $k \in \mathbf{F}_p$. Now, for $x \in \mathbf{F}_p$, Remark (VII) yields

$$\mathbf{N}((1-a)x + 1) - 1 = \prod_{i=0}^{q-1} ((1-a)x + 1)^{p^i} - 1$$

$$= \prod_{i=0}^{q-1} ((1-a)^{p^i} x + 1) - 1$$

$$= \mathbf{N}(1-a)x^q + \cdots + \mathrm{Tr}(1-a)x$$

and every element $k \in \mathbf{F}_p$ is a solution to this polynomial. Since $a \neq 1$ we know $\mathbf{N}(1-a) \neq 0$. Thus the polynomial above is of degree q and has p distinct solutions. This yields $p \leq q$. □

Lemma C.2. $|E| \geq 2$.

Note. The proof requires only that p and q be odd primes that satisfy (A).

Proof. First suppose that $q \geq 5$ and let $s \in P^{\#}$. Since P is abelian, the conjugacy class in H of any element $s \in P$ is simply $\{ s^u \mid u \in U \}$. Let K_i be the conjugacy class of s^i in H and $\hat{K}_i$ the corresponding class sum. The conjugacy class sums form a basis of the center of the group algebra of H, so we can define e to be the coefficient of $\hat{K}_2$ in the product $\hat{K}_1 \hat{K}_1$. It is easy to check (see the discussion in **G** on pages 126-7) that,

$$e = \mathrm{card}\left\{ (s^u, s^v) \mid u, v \in U \text{ and } s^u s^v = s^2 \right\}.$$

The condition $s^u s^v = s^2$ means that $us + vs = 2s$ in $\mathbf{F}_{p^q}$, so $u = 2 - v$ and hence $v \in E$. As a result $e = |E|$.

Now, as noted in (IX), H is a Frobenius group. By Theorem 13.8, p. 68 of [4] (or by applying **G**, Theorem 4.2.1(i), p. 119 and Theorem 4.5.3, p. 143), the irreducible characters of H consist of $|U|$ linear characters with P in their kernel and $(|P| - 1)/|U|$ characters of degree $|U|$ induced from P. Therefore

$$\frac{|P| - 1}{|U|} = \frac{p^q - 1}{(p^q - 1)/(p-1)} = p - 1.$$

Denote by χ_i the $p-1$ characters of degree $|U|$.

By **G**, Theorem 4.2.12, p. 128,

(C.1)

$$e = \frac{|K_1|^2}{|H|} \sum_{\chi \in \mathrm{Irr}(H)} \frac{\chi(s)^2 \overline{\chi(s^2)}}{\chi(1)} = \frac{|U|}{p^q} \left(|U| + \frac{1}{|U|} \sum_{i=1}^{p-1} \chi_i(s)^2 \overline{\chi_i(s^2)} \right).$$

By the orthogonality relations, for every natural number j not divisible by p,

$$\sum_{i=1}^{p-1} |\chi_i(s^j)|^2 \leq \sum_{\chi \in \mathrm{Irr}(H)} |\chi(s^j)|^2 = |C_H(s^j)| = |P| = p^q.$$

Rewriting (C.1), we obtain

$$|p^q e - |U|^2| = \left| \sum_{i=1}^{p-1} \chi_i(s)^2 \overline{\chi_i(s^2)} \right| \leq \left(\max_i |\chi_i(s^2)| \right) \left(\sum_{i=1}^{p-1} |\chi_i(s)|^2 \right)$$

$$\leq \left(\sum_{i=1}^{p-1} |\chi_i(s^2)|^2 \right)^{1/2} \left(\sum_{i=1}^{p-1} |\chi_i(s)|^2 \right) \leq p^{3q/2}.$$

Thus

$$e \geq p^{-q} \left(|U|^2 - p^{3q/2} \right).$$

But $|U| = \dfrac{p^q - 1}{p - 1} > p^{q-1}$ and we have assumed $q \geq 5$, so finally

$$|E| = e \geq p^{q-2} - p^{q/2} > 1.$$

It remains to show that $|E| \geq 2$ when $q = 3$. Suppose for any $c \in \mathbf{F}_p$ that the polynomial

$$f_c(x) = x(x - 2)(x - c) + (x - 1)$$

has a root in $\mathbf{F}_p$. Clearly, for every c, we have $f_c(0) \neq 0$ and $f_c(2) \neq 0$. Thus there must exist distinct elements a and $b \in \mathbf{F}_p$ such that $f_a(x)$ and $f_b(x)$ have a common root $d \neq 0, 2$. But then $d(d - 2)(d - a) = d(d - 2)(d - b)$, which implies that $a = b$, a contradiction.

Thus there is an element $c \in \mathbf{F}_p$ such that $f_c(x)$ has no root in $\mathbf{F}_p$. Suppose a is a root of $f_c(x)$ in $\mathbf{F}_{p^3}$ and let $\mathrm{Gal}(\mathbf{F}_{p^3}/\mathbf{F}_p) = \langle \alpha \rangle$. Then

$$f_c(x) = x(x - 2)(x - c) + (x - 1) = (x - a)(x - a^\alpha)(x - a^{\alpha^2})$$

$$f_c(0) = -1 = (-a)(-a^\alpha)(-a^{\alpha^2}) = -\mathbf{N}(a)$$

$$f_c(2) = 1 = (2 - a)(2 - a^\alpha)(2 - a^{\alpha^2})$$

$$= (2 - a)(2 - a)^\alpha (2 - a)^{\alpha^2} = \mathbf{N}(2 - a).$$

It follows that $a \in E$ and, since $1 \in E$ as well, we have $|E| \geq 2$. $\quad\square$

By Lemmas C.1 and C.2, to prove Theorem C it suffices to show that $E = E^{-1}$. We do this now.

Lemma C.3. $E = E^{-1}$.

Note. Lemma C.3 is easy to prove for $p = 3$. If $\mathbf{N}(a) = \mathbf{N}(2 - a) = 1$, then $\mathbf{N}(a^{-1}) = 1$ and $\mathbf{N}(2 - a^{-1}) = \mathbf{N}(a^{-1}(2a - 1)) = \mathbf{N}(a^{-1}(2 - a)) = 1$. The work mentioned in (IV) shows that whenever p and q are primes that satisfy (A) and $E = E^{-1}$, then $p \leq 3$.

Proof. Let $s \in P_0^{\#}$, $t = s^y$, and $P_1 = P_0^y$.

Step 1. For every $x \in PU$ there exist $u, v \in U$ and $s_1 \in P_0$ such that $x = us_1v$.

Proof. If $x \in PU$, then $x = s'u'$, for some elements $s' \in P$ and $u' \in U$. If $s' \neq 1$, then $s' \in \mathbf{F}_{pq}^* = \mathbf{F}_p^* \times U$ by (VIII). Hence there is an $s_1 \in P_0$ and $u \in U$ such that $s' = us_1u^{-1}$, so $x = s'u' = us_1u^{-1}u' = us_1v$ for $v = u^{-1}u' \in U$. $\square$

Step 2. Let $s_1, s_2 \in P_0$ and $u \in U$. Then $s_1us_2 \in U$ implies either

(1) $s_1 = s_2 = 1$ or
(2) $u = 1$ and $s_1s_2 = 1$.

Proof. If $s_1 \neq 1$, then $s_2 \neq 1$. Since $s_1us_2 = s_1s_2^{u^{-1}}u \in U$, we have $s_1s_2^{u^{-1}} = 1$, or $s_1 + s_2/u = 0$ when considered as elements of $\mathbf{F}_{pq}$. Then $u = -s_2/s_1$, so $u \in U \cap F_p^* = 1$ and $s_1 + s_2 = 0$ in $\mathbf{F}_{pq}$, i.e., $s_1s_2 = 1$ in P_0. $\square$

Step 3. If $t_1 \in P_1^{\#}$, then $(PU) \cap (PU)^{t_1} = U$.

Proof. Let $X = (PU) \cap (PU)^{t_1}$. Since P_1 normalizes U, we have $U \subseteq X$. If $x = s'u'$, where $s' \in P$ and $u' \in U$, then $s' = x(u')^{-1} \in P \cap X$. Thus $X \subseteq (P \cap X)U$ and hence $X = (P \cap X)U$.

Suppose that $X \neq U$. Since U operates irreducibly on P by (VIII), it follows that $X = PU$ and hence t_1 normalizes PU. Since P char PU, we know t_1 normalizes P and therefore $\langle t_1 \rangle = P_1$ also normalizes P. Thus P_1 normalizes $P \cap QP_0 = P_0$. Since $|QP_0|_p = p$, it follows that P_0 and P_1 are Sylow p-subgroups of QP_0 and hence $P_0 = P_1$. But then P_0 normalizes U and $[P_0, U] \subseteq U \cap P = 1$, and finally U centralizes P_0, a contradiction. $\square$

Step 4. Conclusion.

Proof. Let $a \in E^{\#}$ and let b be the element of E such that $a + b = 2$ in $\mathbf{F}_{pq}$. Then $as + bs = 2s$ so in G we have $s^as^b = s^2$, and

(C.2)
$$s^{-2}a^{-1}sab^{-1}sb = 1.$$

Thus, if $k \in \mathbf{F}_p$ and $\ell = k - 2$, after multiplying on the left by $t^{-k+2} = t^{-\ell}$ and on the right by $t^{k-2} = t^\ell$ we have

(C.3)

$$t^{-\ell} \overbrace{s^\ell s^{-k}}^{s^{-2}} t^k \overbrace{(a^{-1})^{t^k}}^{a^{-1}} \overbrace{t^{-k} s^k}^{s} s^{-k+1} t^{k-1} \overbrace{(ab^{-1})^{t^{k-1}}}^{ab^{-1}} \overbrace{t^{-k+1} s^{k-1}}^{s} s^{-\ell} t^\ell \overbrace{b^{t^\ell} t^{-\ell}}^{b} t^\ell = 1.$$

$$\underbrace{s^{-k} t^k}_{} t^{-\ell} s^\ell \qquad \underbrace{s^{-k+1} t^{k-1}}_{} t^{-k} s^k \qquad \underbrace{s^{-\ell} t^\ell}_{} t^{-k+1} s^{k-1}$$

Now observe that

$$s^{-i} t^i = s^{-i}(s^y)^i = s^{-i}(y^{-1}sy)^i = [s^i, y] \in Q.$$

Since Q is commutative, (C.3) becomes

(C.4)

$$s^{-k} t^2 \underbrace{s^{k-2}(a^{-1})^{t^k} s^{-k+1}}_{u_1 s_1 v_1} t^{-1} \underbrace{s^k (ab^{-1})^{t^{k-1}} s^{-k+2}}_{u_2 s_2 v_2} t^{-1} \underbrace{s^{k-1} b^{t^{k-2}} s^{-k}}_{u_3 s_3 v_3} s^k = 1.$$

By Step 1, there are elements u_i, and $v_i \in U$ and $s_i \in P_0$, $(1 \le i \le 3)$, such that

(C.5)

$$u_1 s_1 v_1 = s^{k-2}(a^{-1})^{t^k} s^{-k+1}$$
$$u_2 s_2 v_2 = s^k (ab^{-1})^{t^{k-1}} s^{-k+2}$$
$$u_3 s_3 v_3 = s^{k-1} b^{t^{k-2}} s^{-k}$$

and by Steps 2 and 3

(C.6) $$s_i \ne 1 \qquad (i = 1, 2, 3).$$

If we multiply equation (C.4) on the left by s^k and on the right by s^{-k} and use equation (C.5) we have

$$t^2 u_1 s_1 v_1 t^{-1} u_2 s_2 v_2 t^{-1} u_3 s_3 v_3 = 1, \text{ and hence}$$

$$u_1^{t^{-2}} t^2 s_1 \underbrace{v_1 u_2^t}_{w_3} t^{-1} s_2 t^{-1} \underbrace{v_2^{t^{-1}} u_3}_{w_1} s_3 \underbrace{v_3 u_1^{t^{-2}}}_{w_2} (u_1^{-1})^{t^{-2}} = 1.$$

If we set

$$w_1 = v_2^{t^{-1}} u_3, \qquad w_2 = v_3 u_1^{t^{-2}}, \qquad \text{and} \qquad w_3 = v_1 u_2^t,$$

then $w_i \in U$ and

(C.7) $$t^{-1} s_2 t^{-1} = (w_1 s_3 w_2 t^2 s_1 w_3)^{-1}.$$

Next we show that (C.5) holds with a, b, u_i, and v_i replaced by a^p, b^p, u_i^p, and v_i^p, respectively. We prove only the first equation since the proofs of all three are similar. First observe that in $\mathbf{F}_{p^q}$, $a^p + b^p = (a + b)^p = 2^p = 2$

so that $a^p \in E$. Regarding the first equation of (C.5) modulo P, since $U \cap P = 1$ we see that $(a^{-1})^{t^k} = u_1 v_1$. Hence

$$s_1^{u_1^{-1}} u_1 v_1 = u_1 s_1 v_1 = s^{k-2}(a^{-1})^{t^k} s^{-k+1}$$

$$= s^{k-2} u_1 v_1 s^{-k+1}$$

$$= s^{k-2}(s^{-k+1})^{v_1^{-1} u_1^{-1}} u_1 v_1$$

and so

(C.8)
$$s_1^{u_1^{-1}} = s^{k-2}(s^{-k+1})^{v_1^{-1} u_1^{-1}}, \text{ and}$$

$$s_1 = (s^{k-2})^{u_1}(s^{-k+1})^{v_1^{-1}}.$$

It is enough to show that this equation holds with u_1 and v_1 replaced by u_1^p and v_1^p, respectively.

Writing (C.8) in $\mathbf{F}_{p^q}$ gives $s_1 = (k-2)su_1 + (-k+1)s/v_1$. Taking p^{th} powers, we get $s_1 = (k-2)su_1^p + (-k+1)s/v_1^p$, which implies that in G

$$s_1 = (s^{k-2})^{u_1^p}(s^{-k+1})^{v_1^{-p}},$$

as desired.

It now follows that (C.6) is still true with w_i^p in place of w_i for ($i = 1, 2, 3$). Hence

$$w_1 s_3 w_2 t^2 s_1 w_3 = w_1^p s_3 w_2^p t^2 s_1 w_3^p,$$

which implies

(C.9) $\quad t^{-2} w_2^{-p} s_3^{-1} w_1^{1-p} s_3 w_2 t^2 = s_1 w_3^{p-1} s_1^{-1} \in (PU) \cap (PU)^{t^2}.$

Hence, by Step 3, $s_1 w_3^{p-1} s_1^{-1} \in U$. Since $s_1 \neq 1$, it follows from Step 2 that $w_3^{p-1} = 1$ and hence $w_3 = 1$ by (A). Equation (C.9) then shows that $s_3^{-1} w_1^{1-p} s_3 = w_2^{p-1} \in U$, so Step 2 gives $w_1^{1-p} = 1 = w_2^{p-1}$ and hence $w_1 = w_2 = w_3 = 1$. Then (C.7) becomes $t^{-1} s_2 t^{-1} = s_1^{-1} t^{-2} s_3^{-1}$, i.e.,

(C.10) $\qquad t^2 s_1 t^{-1} s_2 t^{-1} s_3 = 1$

Regarding (C.10) modulo Q, since $P_0 \cap Q = 1$ we see that $s_1 s_2 s_3 = 1$. Hence (C.10) becomes

$$\underbrace{y^{-1} s^2 y}_{t^2} \underbrace{s^{-2} s^2 s_1}_{s_1} \underbrace{y^{-1} s_1^{-1} s^{-2} s s_1 y}_{t^{-1}} \underbrace{s_1^{-1} s^{-1} s s_3^{-1}}_{s_2} \underbrace{y^{-1} s_3 s^{-1} s_3^{-1} y}_{t^{-1}} \underbrace{s_3}_{s_3} = 1.$$

Identifying P_0 with its image in $\text{End}([Q, P_0])$ and regrouping, we have

$$y^{-1} y^{s^{-2}} y^{-s_1^{-1} s^{-2}} y^{s_1^{-1} s^{-1}} y^{-s_3 s^{-1}} y^{s_3} = 1.$$

Thus we see that y is in the kernel of the map

$$-1 + s^{-2} - s_1^{-1} s^{-2} + s_1^{-1} s^{-1} - s_3 s^{-1} + s_3 = (s^{-1} + 1 - s_1^{-1} s^{-1} - s_3)(s^{-1} - 1).$$

If $x \in [Q, P_0]$ is chosen such that $x^{s^{-1}} = x$, then x centralizes s and hence centralizes P_0. Then $x \in C_Q(P_0) \cap [Q, P_0] = 1$ by (X). Thus s^{-1} operates without fixed points on $[Q, P_0]$, so y is in the kernel of $s^{-1} + 1 - s_1^{-1} s^{-1} - s_3$. Therefore we have

$$y^{-s_3 + s^{-1} - s_1^{-1} s^{-1} + 1} = y^{-s_3} y^{s^{-1}} y^{-s_1^{-1} s^{-1}} y = 1.$$

Setting $t_i = y^{-1} s_i y$ for $i = 1, 2$, and 3, it follows that

$$s_3^{-1} t_3 t s_1 t_1^{-1} t^{-1} = s_3^{-1} \overbrace{y^{-1} s_3 s y}^{t_3 t} \overbrace{s^{-1} s s_1}^{s_1} \overbrace{y^{-1} s_1^{-1} s^{-1} y}^{t_1^{-1} t^{-1}}$$

$$= y^{-s_3} y^{s^{-1}} y^{-s_1^{-1} s^{-1}} y = 1$$

so $s_1 t_1^{-1} t^{-1} = t^{-1} t_3^{-1} s_3$.

It now follows that if $u \in U^{\#}$, then

$$u^{s_1 t_1^{-1} t^{-1}} = u^{t^{-1} t_3^{-1} s_3} \in (PU) \cap (PU)^{t_1^{-1} t^{-1}}.$$

If $t_1 \neq t^{-1}$, then by Step 3, $u^{s_1 t_1^{-1} t^{-1}} \in U$, and so $u^{s_1} \in U^{t t_1} = U$ since $t t_1 \in P_1$. But then $u^{s_1} = s_1^{-1} u s_1 \in U$ implies, by Step 2, that $s_1 = 1$, contradicting (C.6). Hence $t_1 = t^{-1}$ and $s_1 = s^{-1}$.

For $k = 3$ the first equation of (C.5) gives us

$$s(a^{-1})^{t^3} s^{-2} = u_1 s^{-1} v_1, \text{ and hence}$$

$$s^2 = v_1^{-1} s u_1^{-1} s(a^{-1})^{t^3}.$$

Regarding this equation modulo P, we get $v_1^{-1} u_1^{-1} (a^{-1})^{t^3} = 1$, i.e., $v_1 = u_1^{-1} (a^{-1})^{t^3}$. Then

$$s^2 = v_1^{-1} s u_1^{-1} (a^{-1})^{t^3} s^{(a^{-1})^{t^3}} = s^{v_1} s^{(a^{-1})^{t^3}},$$

and so $(v_1 + (a^{-1})^{t^3}) s = 2s$ in $\mathbf{F}_{p^q}$. This yields $v_1 = 2 - (a^{-1})^{t^3}$, and $\mathbf{N}(2 - (a^{-1})^{t^3}) = \mathbf{N}(v_1) = 1$. Since $(a^{-1})^{t^3} \in U$, it follows that $(a^{-1})^{t^3} \in E$.

We have now shown that if $a \in E$ then $(a^{-1})^{t^3} \in E$. By induction it follows that

$$\left(a^{(-1)^n}\right)^{t^{3n}} \in E$$

for any natural number n. Taking $n = p$, we get $a^{-1} \in E$ as desired. $\square$

Problem 1. (See (IV).) Can $p = 3$ in Theorem C?

APPENDIX D

CN-Groups of Odd Order

The proof in **G** that every CN-group of odd order is solvable requires extensive passages in Chapters 7, 8, and 10 of **G**. While this material is worthwhile for a deeper understanding of group theory, most of it is unnecessary for the CN-theorem if one combines ideas from Gorenstein's proof with ideas from the proof in W. Feit's *Characters of Finite Groups* [6]. We indicate how to do this now.

One first reads Chapters 1–8 of **G** or the substitute prerequisites that are described in Appendix A, as well as Sections 1–4 of this work and Lemma 10.1.3 of **G**. Then one notes that for G solvable of odd order, Theorems 7.6.1 and 10.2.1 of **G** follow from our Theorems 4.18(b) and 3.7, while the proofs of Theorems 7.6.2 and 10.3.1 reduce to one paragraph and one sentence respectively. One proceeds to the introduction of Chapter 14 and to Section 14.1, which is slightly easier for G of odd order (but not necessarily solvable). Lemma 14.2.1 is easy, but then it is useful to insert the following lemma, suggested by the proof of (27.6) in [6].

Lemma D.1. Suppose G is a minimal simple CN-group of odd order, p is a prime, P and Q are Sylow p-subgroups of G, and $P \cap Q \neq 1$. Then $P = Q$.

Proof. Assume the result is false. We will obtain a contradiction. Take P to violate the conclusion for some Q. Let $N = N_G(Z(J(P)))$. (One may substitute $L(P)$ for $J(P)$ throughout this proof if one prefers to use Theorem B.4 instead of Theorem 6.2.)

Take any Q such that $P \cap Q$ is maximal subject to P and Q violating the conclusion of the theorem. Let M be a subgroup of G maximal subject to containing $N_G(P \cap Q)$ and satisfying $\mathcal{O}_p(M) \neq 1$. Let P_1 and Q_1 be Sylow p-subgroups of G containing $N_P(P \cap Q)$ and $N_Q(P \cap Q)$, and let P_2 be a Sylow p-subgroup of G containing P_1. Since $P \cap Q \subset P$,

$$P \cap Q \subset N_P(P \cap Q) \subseteq P \cap P_1 \subseteq P \cap P_2.$$

By the choice of P and Q, we have $P_2 = P$, and hence $P_1 \subseteq P$. Similarly, $Q_1 \subseteq Q$. Thus

(D.1)
$$P \cap Q \subset P \cap M \text{ and}$$
$$P \cap M \text{ and } Q \cap M \text{ are Sylow } p\text{-subgroups of } M.$$

Since $M \subset G$, we have M solvable. Moreover, $\mathcal{O}_p(M) \neq 1$. By (D.1),

$$(P \cap M) \cap (Q \cap M) = P \cap Q \subset P \cap M, \text{ and hence } P \cap M \neq Q \cap M.$$

Therefore $P \cap M \neq \mathcal{O}_p(M)$ or $Q \cap M \neq \mathcal{O}_p(M)$ (actually both). By Corollary 14.1.6 of **G**, M is a 3-step group with respect to p. From the definition of a 3-step group and a short argument $\mathcal{O}_{p'}(M) = 1$ and

(D.2) $$P \cap Q = \mathcal{O}_p(M).$$

Hence, by Theorem 6.2,

$$Z(J(P)) = Z(J(P))\mathcal{O}_{p'}(M) \lhd M.$$

So $M \subseteq N$. By the maximal choice of M, we have $M = N$. Furthermore, by the definition of a 3-step group,

$$N/\mathcal{O}_{p,p'}(N) \text{ is a nonidentity } p\text{-group.}$$

Let $K = \mathcal{O}_{p,p'}(N)$. Then $N/K \supset (N/K)' = N'K/K$, and the quotient is an abelian p-group. So

(D.3) $$P \cap N'K \subset P.$$

We wish to apply the Focal Subgroup Theorem (Theorem 1.17) to P and G. Suppose x, $y \in P$ and x is conjugate to y in G. We claim that $x^{-1}y \in P \cap N'K$. We may assume that x, $y \neq 1$. Take $t \in G$ such that $x^t = y$. If $P^t = P$, then

$$t \in N(P) \subseteq N(Z(J(P))) \subseteq N \text{ and}$$
$$x^{-1}y = x^{-1}t^{-1}xt \in P \cap N' \subseteq P \cap N'K,$$

as desired. If $P^t \neq P$, take a Sylow p-subgroup Q of G such that Q is maximal subject to

$$Q \neq P \text{ and } Q \cap P \supseteq P^t \cap P.$$

Then $y \in P \cap Q$ and, by (D.2),

$$y \in P \cap Q = \mathcal{O}_p(N) \subseteq \mathcal{O}_{p,p'}(N) = K.$$

Similarly, since $x = y^{t^{-1}}$, $x \in K$. Thus $x^{-1}y \in P \cap K \subseteq P \cap N'K$, as desired. This proves the claim, and (D.3) and the Focal Subgroup Theorem yield

$$P \cap G' = \left\langle x^{-1}y \mid x, y \in G \text{ and } x \text{ and } y \text{ are conjugate in } G \right\rangle$$
$$\subseteq P \cap N'K \subset P.$$

But $G' = G$, because G is simple, a contradiction. $\square$

Lemma D.1 has some easy consequences.

Lemma D.2. Suppose P is a nonidentity Sylow subgroup of a minimal simple CN-group G of odd order and $N = N_G(P)$. Then $P \subseteq N'$.

Proof. Suppose $x, y \in P$ and $t \in G$ and $x^t = y$. If $x = y = 1$, then $x^{-1}y \in N'$. Otherwise, $P \cap P^t \neq 1$, and then $P = P^t$ by Lemma D.1, whence

$$t \in N \text{ and } x^{-1}y = x^{-1}t^{-1}xt \in P \cap N'.$$

By the Focal Subgroup Theorem, $P \cap G' \subseteq P \cap N'$. Since $G' = G$, we have $P \subseteq N'$. $\square$

It is easy to see from Lemma D.1 that a minimal simple CN-group of odd order cannot contain a 3-step group as a subgroup, which yields Theorem 14.2.2. In addition, Lemmas D.1 and D.2 simplify the proof of Theorem 14.2.3; one may use $N_G(P)$ in place of $N_G(Z(J(P)))$, since $N_G(P)$ is not nilpotent by Lemma D.2. The rest of the proof of the CN-theorem may be read as in **G** without change.

APPENDIX E

Further Results of Feit and Thompson

In Theorem 15.8 and Corollary 15.9, we presented some new results of Feit and Thompson about the situation in which

$$x \in M_\sigma, \quad C_G(x) \not\subseteq M, \quad N \in \mathcal{M}(C_G(x)), \quad \text{and} \quad N \in \mathcal{M}_{\mathscr{P}_2}.$$

Here we present some additional results of theirs and some applications that shed further light on this situation and could lead to further reductions in the proof of the Odd Order Theorem.

The following result was proved by Philip Hall for applications to his theory of regular p-groups, using his commutator collecting process. It may be found on pp. 37–41 of [26] and in many other books (e.g., [17, pp. 315–318]).

Theorem E.1. Let

$$G = G_1 \supseteq G_2 \supseteq G_3 \supseteq \cdots$$

be the lower central series of a group G. Take $x, y \in G$ and a positive integer n. For $r = 2, 3, \ldots, n$, let e_r be the usual binomial coefficient

$$\binom{n}{r} = \frac{n(n-1)(n-2)\cdots(n-r+1)}{r!}.$$

Then there exist elements $c_r \in G_r$ for $r = 2, 3, \ldots, n$ such that

$$x^n y^n = (xy)^n c_2^{e_2} \cdots c_n^{e_n}.$$

Proposition E.2. Suppose p is a prime and R is a p-group of nilpotence class at most $p - 1$. Define a mapping ϕ of R into R by $\phi(x) = x^p$. Then

(a) $\Omega_1(R)$ has exponent 1 or p, and
(b) if $R' \subseteq \Omega_1(R)$, then ϕ is a homomorphism.

Proof. This result follows immediately from Philip Hall's theory of regular p-groups (e.g., [26, pp. 41–49]). Since it is easy to derive from Theorem E.1, we will do so.

Step 1. Suppose R' has exponent 1 or p. Then ϕ is a homomorphism.

Proof. Take any x, $y \in R$. We want to show that $(xy)^p = x^p y^p$. Apply Theorem E.1 with $n = p$. Since G has nilpotence class at most $p - 1$, we have $c_p \in G_p = 1$. For $r = 2, \ldots, p - 1$, $c_r \in R'$ and $e_r = \binom{p}{r}$, which is divisible by p, whence $c_r^{e_r} = 1$. Therefore, $x^p y^p = (xy)^p$. $\square$

Step 2. Conclusion.

Proof. Clearly, (b) will follow from (a) and Step 1. We prove (a).

Use induction on $|R|$. It is sufficient to show that $(xy)^p = 1$ for any x, $y \in R$ of order 1 or p.

We may assume that $\langle y \rangle \subset R = \langle x, y \rangle$. Take a maximal subgroup M of R that contains $\langle y \rangle$. Then $M \lhd R$. By induction $\Omega_1(M)$ has exponent p. Since

$$\Omega_1(M) \lhd R \text{ and } R = \langle x, y \rangle = \langle x, \Omega_1(M) \rangle,$$

$R/\Omega_1(M)$ is cyclic and $R' \subseteq \Omega_1(M)$. Then, by Step 1,

$$(xy)^p = \phi(xy) = \phi(x)\phi(y) = x^p y^p = 1. \quad \square$$

Recall (from Section 1) an operator group A acts *regularly* on G if $C_G(\alpha) = 1$ for all $\alpha \in A^{\#}$.

Theorem E.3 (Feit and Thompson, 1991). Suppose p and q are distinct odd primes, R is a p-group, R_0 and R_1 are nonidentity subgroups of R, B is an operator group on R, and A is a subgroup of B. Assume that p does not divide $|B|$ and

$$|A| = q, \ |R_0| = p, \ R_1 \text{ is cyclic}, \ C_R(R_0) = R_0 \times R_1,$$

and A fixes R_0 and acts regularly on R. Then

(a) q divides $\frac{1}{2}(p - 1)$,
(b) $\Omega_1(R)$ has exponent p, $R_0 \not\subseteq (\Omega_1(R))'$, and $|\Omega_1(R)/(\Omega_1(R))'| = p^2$,
(c) $|\Omega_1(R)| \leq p^q$, and
(d) if B fixes $R_0\Phi(\Omega_1(R))$, then B fixes R_0.

Proof. Step 1. Part (a) is valid and $p \geq 7$.

Proof. Since A acts regularly on R, it acts regularly on R_0. Consequently $q \mid (p - 1)$. Since p and q are odd, we obtain (a), and $p \geq 2q + 1 \geq 7$. $\square$

Step 2. Let S be any A-invariant subgroup of R of exponent p that properly contains R_0. Then $R_0 \not\subseteq S'$, $|S| \leq p^q$, and $|S/S'| = p^2$.

Proof. Since $q \geq 3$, an examination of the p-groups of order at most p^3 yields the conclusion when $|S| \leq p^3$, so we will assume that

(E.1) $$|S| \geq p^4.$$

Take $V \in \mathrm{SCN}(S)$. Then

(E.2) $$|S/V| \text{ divides } |\mathrm{Aut}(V)| \text{ and } |S| = |S/V||V|.$$

Since S has exponent p, V is elementary abelian. If $|V| \leq p^2$, then (E.2) gives

$$|S/V| \leq p \text{ and } p^4 \leq |S| \leq pp^2 = p^3,$$

a contradiction. Thus $|V| \geq p^3$ and

(E.3) $$r(S) \geq m(V) \geq 3.$$

Let $Z = \Omega_1(Z(S))$. Since $C_R(R_0) = R_0 \times R_1$, we have $R_0 \cap Z = 1$. Clearly,

$$R_0 \times Z \subseteq C_S(R_0) \subseteq R_0 \times \Omega_1(R_1),$$

so that

(E.4) $$|Z| = p \text{ and } C_S(R_0) = R_0 \times Z.$$

Note that S is narrow. Let $E = C_S(R_0)$ and take T as in Lemma 5.2 (with S in place of R). Then, by Lemma 5.2 and Theorem 5.3(d),

(E.5) $$T \text{ char } S, \ |S : T| = |C_T(R_0)| = p, \text{ and } R_0 \cap T = 1.$$

Since $S' \subseteq T$, we have $R_0 \not\subseteq S'$ and $S = R_0 T$. We now follow the part of the proof of Theorem 5.5 that comes after (5.5). We have an A-invariant series of subgroups

$$T = H_0 \supset H_1 \supset \cdots \supset H_n = 1,$$

for which

(E.6)
$$H_i = [R, H_{i-1}] = [R_0, H_{i-1}] \text{ and } |H_{i-1} : H_i| = p,$$
$$\text{for } i = 1, 2, \ldots, n.$$

Thus $|T| = p^n$.

Let

$$S = S^1 \supseteq S^2 \supseteq S^3 \supseteq \cdots$$

be the lower central series of S. Since

$$|H_0/H_1| = p \text{ and } H_1 = [R_0, H_0] = [R_0, T] \lhd R_0 T = S,$$

we have $|S/H_1| = p^2$ and $S^2 = [S, S] \subseteq H_1 = [R_0, T] \subseteq S^2$. Thus

(E.7) $$H_1 = S^2 \text{ and } |S/S'| = |S/S^2| = p^2.$$

Similarly, by induction, we see that

(E.8) $\qquad\qquad\qquad H_i = S^{i+1} \lhd S,\ \text{for}\ i = 1, 2, \ldots, n.$

Let $v \in R_0^\#$, $\alpha \in A^\#$, and $w \in H_0 - H_1$ and define

$$w_0 = w\ \text{and}\ w_i = [w_{i-1}, v],\ \text{for}\ i = 1, 2, \ldots, n - 1.$$

Then $v^\alpha = v^r$ for some integer r such that $r^q \equiv 1 \pmod{p}$. Similarly, by (E.6), for $i = 0, 1, \ldots, n - 1$,

(E.9)
$$w_i{}^\alpha \equiv w_i{}^{r_i} \pmod{H_{i+1}}$$
$$\text{for some integers}\ r_i\ \text{such that}\ r_i{}^q \equiv 1 \pmod{p}.$$

Moreover, if $r_i \equiv 1 \pmod{p}$ for some i, then A centralizes H_i/H_{i+1} and, by Proposition 1.5(d),

$$H_i = C_{H_i}(A)H_{i+1},$$

contrary to the regular action of A on R. Thus

(E.10) $\qquad\qquad\qquad r_i \not\equiv 1 \pmod{p},\ \text{for}\ i = 0, 1, \ldots, n - 1.$

Similarly

(E.11) $\qquad\qquad\qquad\qquad r \not\equiv 1 \pmod{p}.$

Now take any integer i such that $1 \le i \le n - 1$. Let $\overline{S} = S/H_{i+1}$ and apply the bar convention. Then $|\overline{H}_i| = p$ and $\overline{H}_i \lhd \overline{S}$. So

$$\langle \overline{w}_i \rangle = \overline{H}_i \subseteq Z(\overline{S})\ \text{and}$$
$$\overline{H}_i = [\overline{R}_0, \overline{H}_{i-1}] = [\langle \overline{v} \rangle, \langle \overline{w}_{i-1}, \overline{H}_i \rangle] = [\langle \overline{v} \rangle, \langle \overline{w}_{i-1} \rangle].$$

By (E.9), $w_{i-1}{}^\alpha = w_{i-1}{}^{r_{i-1}}u$ for some $u \in H_i$. By Lemma 4.2(a) (applied to $\langle \overline{R}_0, \overline{H}_{i-1} \rangle$), we have

$$w_i{}^{r_{i-1}r} \equiv [w_{i-1}, v]^{r_{i-1}r} \equiv [w_{i-1}{}^{r_{i-1}}, v^r] \equiv [w_{i-1}{}^{r_{i-1}}u, v^r]$$
$$\equiv [w_{i-1}{}^\alpha, v^\alpha] \equiv [w_{i-1}, v]^\alpha \equiv w_i{}^\alpha \equiv w_i{}^{r_i} \pmod{H_{i+1}}.$$

Hence

$$r_i \equiv r_{i-1}r \pmod{p}.$$

We now see by induction that, for $i = 0, 1, \ldots, n - 1$,

(E.12) $\qquad\qquad\qquad\qquad r_i \equiv r_0 r^i \pmod{p}.$

Recall that the nonzero integers $\pmod{p}$ form a cyclic group of order $p - 1$ and $r^q \equiv 1 \pmod{p}$. Therefore, by (E.10) and (E.11), there exists an integer j such that $1 \le j \le q - 1$ and $r^j \equiv r_0 \pmod{p}$. By (E.12), none of the integers $r^j, r^{j+1}, \ldots, r^{j+n-1}$ is congruent to 1 $\pmod{p}$. Therefore

$$q - 1 \ge j + n - 1 \ge n\ \text{and}\ |S| = |R_0 T| = pp^n \le p^q.$$

This completes the proof of Step 2. $\qquad\square$

Step 3. Parts (b) and (c) are valid.

Proof. Take an A-invariant subgroup S of R of exponent p that is maximal subject to containing $R_0 \times \Omega_1(R_1)$. Then $S \subseteq \Omega_1(R)$. By Step 2,

$$(\text{E.13}) \qquad\qquad R_0 \not\subseteq S', \; |S| \leq p^q \text{ and } |S/S'| = p^2.$$

Let $P = \Omega_1(R)$ and $T = N_P(S)$. If $S = \Omega_1(T)$, then

$$N_P(T) \subseteq N_P(\Omega_1(T)) = N_P(S) = T,$$

whence $T = P$ and $S = \Omega_1(P) = \Omega_1(\Omega_1(R)) = \Omega_1(R)$, which, by (E.13), yields (b) and (c). So, we assume that

$$(\text{E.14}) \qquad\qquad\qquad S \subset \Omega_1(T).$$

Clearly, $S \supseteq \Omega_1(Z(R))$ and

$$C_S(R_0) = S \cap (R_0 \times R_1) = R_0 \times \Omega_1(Z(R)).$$

Let $v \in R_0^{\#}$ and let K be the conjugacy class of v in S. Then

$$(\text{E.15}) \qquad\quad |K| = |S : C_S(v)| = |S : C_S(R_0)| = |S|/p^2.$$

Let T_1 be the normalizer of the set K in T. Then $S \subseteq T_1$ and the conjugacy class of v in T is the union of $|T : T_1|$ conjugacy classes of S, each having $|S|/p^2$ elements. Since none contains the identity element,

$$(\text{E.16}) \qquad |T : T_1||S|/p^2 \leq |S| - 1 < |S| \text{ and } |T : T_1| < p^2.$$

By an easy variation of the Frattini argument,

$$T_1 = SC_{T_1}(v) = S(T_1 \cap R_0 R_1) = SR_0(T_1 \cap R_1) = S(T_1 \cap R_1),$$

and $T_1/S \cong (T_1 \cap R_1)/(T_1 \cap R_1 \cap S)$. As R_1 is cyclic, so is T_1/S.

By (E.16), T_1/S is a cyclic subgroup of index 1 or p in T/S. Hence, by Lemma 4.5, $|\Omega_1(T/S)| \leq p^2$. Now, by (E.13) and (E.14),

$$|\Omega_1(T)/S| \leq |\Omega_1(T/S)| \leq p^2 \text{ and } |\Omega_1(T)| \leq p^2|S| \leq p^{q+2}.$$

Since $p \geq 7$ and $q \leq (p-1)/2$, by Step 1,

$$q + 2 \leq 2 + ((p-1)/2) = (p+3)/2 < (p+p-2)/2 = p - 1.$$

Therefore $\Omega_1(T)$ has nilpotence class at most $p - 1$. By Proposition E.2, $\Omega_1(\Omega_1(T))$ has exponent p. Since $\Omega_1(\Omega_1(T)) = \Omega_1(T)$, this contradicts the maximal choice of S, by (E.14). $\square$

Step 4. Part (d) is valid.

Proof. Let $S = \Omega_1(R)$ and let G be the semidirect product of S by B. By (b),

(E.17) $$|S/S'| = p^2 \text{ and } |S'| = |S|/p^2.$$

Note that $\Phi(S) = S'$, because S has exponent p.

Assume that B fixes $R_0 S'$. Let $v \in R_0^{\#}$. For each $x \in S$,

$$x^{-1}vx \equiv vv^{-1}x^{-1}vx \equiv v \pmod{S'},$$

so the conjugacy class of v in S is contained in vS'. By (E.15) (in Step 3), it has $|S|/p^2$ elements, and hence is equal to vS', by (E.17). The same is true of $v^2, v^3, \ldots, v^{p-1}$. Thus

every element of the set $R_0 S' - S'$ is conjugate to an element of $R_0^{\#}$.

Since B fixes $R_0 S'$, it follows that, for each $\beta \in B$,

$$R_0^{\beta} = R_0^{x} \text{ for some } x \in S.$$

By a variation of the Frattini argument (similar to that used in the proof of Step 3),

$$SB = SN_G(R_0).$$

By the Schur-Zassenhaus Theorem, $N_G(R_0)$ contains a complement B^* to $N_G(R_0) \cap S$, and $B^* = B^y$ for some $y \in S$. Therefore

$$B \text{ normalizes } R_0^{y^{-1}}.$$

Then A normalizes R_0 and $R_0^{y^{-1}}$. Therefore

$$R_0^{y^{-1}} = (R_0^{y^{-1}})^\alpha = R_0^{(y^\alpha)^{-1}},$$
$$(y^\alpha)^{-1}y \in N_S(R_0), \quad y^\alpha N_S(R_0) = yN_S(R_0),$$

and α (and A) fix the coset $yN_S(R_0)$. Now, $|A| = q$, while $|yN_S(R_0)| = |N_S(R_0)|$, which is a power of p and hence not divisible by q. Therefore A fixes at least one element, z, of $yN_S(R_0)$. Since A acts regularly on R, $z = 1$. Therefore

$$Vy \in yN_S(R_0) \text{ and } R_0^{y^{-1}} = R_0.$$

This completes the proof of Step 4 and of Theorem E.3. $\square$

Proposition E.4. Assume the situation of Theorem E.3 and let $S = \Omega_1(R)$. Suppose $|S| \geq p^4$, B acts regularly on R, and B does not fix R_0.

Then $C_S(Z_2(S))$ is abelian and has index p in S.

Proof. By Theorem E.3,

(E.18)
$$S \text{ has exponent } p, |S/S'| = p^2, |S| \le p^q,$$
$$\text{and } B \text{ does not fix } R_0\Phi(S)$$

Hence $\Phi(S) = S'$, and S/S' is elementary abelian.

Now take n, T, H_i, H_n, v, α, w_i r, and r_i, for all $i = 0, 1, \ldots, n-1$, as in Step 2 of the proof of Theorem E.3. Then

(E.19)
$$T = C_S(Z_2(S)) \operatorname{char} S, |S : T| = p,$$
$$T = H_0 \supset H_1 \supset \cdots \supset H_n = 1,$$
$$H_i = [R, H_{i-1}], |H_{i-1} : H_i| = p, \text{ and}$$
$$H_{i-1} = \langle W_{i-1}, H_i \rangle, \text{ for } i = 1, 2, \ldots, n,$$
$$v \in R_0^{\#}, \langle \alpha \rangle = A, \text{ and } v^\alpha = v^r, \text{ and } w_0^\alpha \equiv w_0^{r_0} \pmod{H_1}.$$

Therefore

$$|T| = p^n, |S| = p^{n+1}, \text{ and } H_i \operatorname{char} S, \text{ for } i = 0, 1, \ldots, n.$$

Assume that T is not abelian. We will work toward a contradiction. Now $\operatorname{Aut}(S/T)$ is abelian because $|S/T| = p$. So B' centralizes S/T. By Proposition 1.5(d), B' centralizes an element of $S - T$. Since B acts regularly on R, we have $B' = 1$ and

(E.20)
$$B \text{ is abelian.}$$

By (E.18), there exists $\beta \in B$ such that β does not fix $R_0\Phi$. Let us regard S/S' as a 2-dimensional vector space over $\mathbf{F}_p$. Then α has eigenvalues r and r_0 on R_0S'/S' and T/S', respectively. Since β centralizes α, β fixes both subspaces if $r \ne r_0$. Therefore $r = r_0$. Since B fixes T/S' and p does not divide $|B|$, B fixes some complement Q/S' of T/S' in S/S'. Let β have eigenvalues t and t_0 on Q/S' and T/S', respectively. If $t = t_0$, then β fixes every 1-dimensional subspace of S/S', including R_0S'/S'. Thus

(E.21)
$$t \ne t_0.$$

By (E.16) and (E.19) in Step 2 of the proof of Theorem E.3,

(E.22)
$$w_i{}^\alpha \equiv w_I{}^{r_i} \pmod{H_{i+1}}$$
$$\text{for } r_i = r_0 r^i \text{ and } i = 0, 1, \ldots, n-1.$$

Similarly one can show that

(E.23)
$$w_i{}^\beta \equiv w_i{}^{t_i} \pmod{H_{i+1}}$$
$$\text{for } t_i = t_0 t^i \text{ and } i = 0, 1, \ldots, n-1.$$

Take k minimal such that T/H_k is not abelian. Then

(E.24)
$$[H_0, H_0]H_k = T'H_k = H_{k-1}.$$

(Actually $T' = H_{k-1}$, but we do not need this.)

Now take i maximal such that $[H_i, T] \not\subseteq H_k$ and j maximal such that $[H_i, H_j] \not\subseteq H_k$. Then, by (E.19),

(E.25) $\qquad 0 \leq j \leq i \leq k - 2$ and $[w_i, w_j] \in H_{k-1} - H_k$.

By (E.22), there exist $x \in H_{i+1}$ and $x' \in H_{j+1}$ such that

$$w_i^\alpha = w_i^{r_i} \text{ and } w_j^\alpha = w_j^{r_j} x',$$

and similarly

$$[w_i, w_j]^\alpha \equiv [w_i, w_j]^{r_{k-1}} \pmod{H_k}.$$

By Lemma 4.2,

$$[w_i, w_j]^\alpha \equiv [w_i^{r_i} x, w_j^{r_j} x'] \equiv [w_i^{r_i}, w_j^{r_j}] \equiv [w_i, w_j]^{r_i r_j} \pmod{H_k}.$$

Therefore, by (E.25), $r_i r_j = r_{k-1}$.

By (E.22), $r_0 r^i r_0 r^j = r_0 r^{k-1}$, so

(E.26) $\qquad\qquad\qquad r_0 r^{i+j} = r^{k-1}.$

By using β instead of α, we obtain similarly

(E.27) $\qquad\qquad\qquad t_0 t^{i+j} = t^{k-1}.$

Now $p^k = |T/H_k| \leq |T| = p^n = |S|/p \leq p^{q-1}$ by Theorem E.3(c), so $k \leq q - 1$. Recalling that $r_0 = r$, we obtain from (E.26)

$$r^{1+i+j} = f^{k-1}, \ r^{j+2} = r^{k-i}, \text{ and } j + 2 \equiv k - i \pmod{q}.$$

Since $0 \leq j \leq i \leq k - 2 \leq q - 3$, we have $0 \leq j + 2 \leq q - 1$ and $0 \leq k - i \leq k \leq q - 1$, so

$$j + 2 = k - i.$$

Now, by (E.27), $t_0 = t^{k-1-i-j} = t$, a contradiction. This completes the proof of the proposition. $\square$

Corollary E.5. Suppose that M is a maximal subgroup of G, x is an element of M_σ of prime order p, $C_G(x) \not\subseteq M$, and $N \in \mathcal{M}(C_G(x))$. Assume that $N \notin \mathcal{M}_{\mathcal{F}}$ and that

(i) $|M/M'|$ is prime, or

(ii) $\Omega_1(\mathcal{O}_p(M))$ has no normal abelian subgroup of index p.

Then every maximal subgroup of G is of Type I or Type II.

Proof. Note that we are in the situation of Corollary 15.9, with $r = p$. Therefore M is a Frobenius group possessing a cyclic Frobenius complement E such that

(E.28)
$$|E \cap N| = |N/N'|, \ N_E(\langle x \rangle) \subseteq E \cap N,$$
$$N \in \mathcal{M}_{\mathcal{P}_2}, \text{ and } p \in \tau_2(N)$$

Hence $\mathcal{O}_p(M)$ is a Sylow p-subgroup of M_σ and of G. By Proposition 16.1, M and N are of Types I and II, respectively.

Let K_1 be a Hall $\kappa(N)$-subgroup of $M \cap N$ and R be a Sylow p-subgroup of $M \cap N$. As in the proof of Corollary 15.9, we see that

(E.29)
$$\begin{aligned} &C_{N_\sigma}(x) > 1, \ M \cap N \text{ is a complement to } N_\sigma \text{ in } N, \\ &|K_1| \text{ is prime, and } R \text{ is contained in an abelian} \\ &\text{normal complement } U_1 \text{ to } K_1 \text{ in } M \cap N. \end{aligned}$$

Hence

(E.30) $\quad R = \mathcal{O}_p(M) \cap (M \cap N) = \mathcal{O}_p(M) \cap N = C_{\mathcal{O}_p(M)}(x).$

Now we show that (ii) yields (i). Assume (ii), and let $S = \Omega_1(\mathcal{O}_p(M))$ and $R_0 = C_R(N_\sigma)$. Since $p \in \tau_2(M)$, $r(R) = 2$. By (ii), $\mathcal{O}_p(M)$ is not abelian. Therefore, by Theorem 12.7 and (E.30),

(E.31)
$$\begin{aligned} &R_0 = \mathcal{O}_p(F(N)) \lhd N, \quad |R_0| = p, \\ &C_{\mathcal{O}_p(M)}(x) = R = R_0 \times (R \cap E_0), \\ &\text{for some complement } E_0 \text{ to } R_0 \text{ in } M \cap N, \end{aligned}$$

and $R \cap E_0$ acts regularly on N_σ. By Proposition 3.9, $R \cap E_0$ is cyclic. By (E.29), $C_{N_\sigma}(x) \neq 1$. Hence

(E.32) $\quad \langle x \rangle = C_R(N_\sigma) = R_0 \lhd N$ and K_1 normalizes R_0.

Since M is a Frobenius group, E acts regularly on $\mathcal{O}_p(M)$. Now, by (E.31) and (E.32), and Theorem E.3, with $\mathcal{O}_p(M)$, K_1, and E in place of R, A, and B,
$$S \text{ has exponent } p.$$
Hence, by (ii), $|S| \geq p^4$. Furthermore, by Proposition E.4, $E = K_1$, which proves (i).

To complete the proof of the corollary, we can now assume (i) and assume that G contains a maximal subgroup N^* that is neither of Type I nor of Type II. Let $K^* = C_{N_\sigma}(K_1)$, $\widehat{Z} = K_1 K^* - K_1 \cup K^*$, $k = |K_1|$, and $k^* = |K^*|$. For each maximal subgroup L of G and each subset T of G, define the sets $\widetilde{L}$ and $\mathscr{C}_G(T)$ as in Section 14, p. 109. By Lemma 14.5,

(E.33) $\quad |\mathscr{C}_G(\widetilde{L})| = (|L_\sigma| - 1)|G : L| = |G| \left(\dfrac{1}{|L : L_\sigma|} - \dfrac{1}{|L|} \right),$

and the sets $\mathscr{C}_G(\widetilde{L})$ are disjoint for nonconjugate maximal subgroups L. By Theorem 14.7(e), (a), and a short argument,

(E.34) $\qquad |\mathscr{C}_G(\widehat{Z})| = \left(1 - \dfrac{1}{k} - \dfrac{1}{k^*} + \dfrac{1}{kk^*} \right) |G|,$

and $\mathscr{C}_G(\widehat{Z})$ is disjoint from $\mathscr{C}_G(\widetilde{L})$ for every maximal subgroup L.

Now let

$$s = \frac{1}{|G|}(|\mathscr{C}_G(\widetilde{M})| + |\mathscr{C}_G(\widetilde{N}^*)| + |\mathscr{C}_G(\widetilde{N})| + |\mathscr{C}_G(\widehat{Z})|).$$

Then $s \leq 1$. By Theorem 14.7, N^* is conjugate to the unique maximal subgroup containing $C_G(K_1)$. By Proposition 14.2(a) and (g), we know that $|N/N_\sigma| = |K_1 V_1| < |M|$ and $|N^*/N^*_\sigma| = |K^*|$. Therefore, by (E.33) and (E.34),

$$s = \left(\frac{1}{k} - \frac{1}{|M|}\right) + \left(\frac{1}{k^*} - \frac{1}{|N^*|}\right) + \left(\frac{1}{|K_1 U_1|} - \frac{1}{|N|}\right) + \left(1 - \frac{1}{k} - \frac{1}{k^*} + \frac{1}{kk^*}\right).$$

Since KK^* is a proper subgroup of N, $|N| \geq 3kk^*$. Similarly $|N^*| \geq 3kk^*$. Since $|K_1 U_1| < |M|$, we obtain

$$s > 1 + \left(\frac{1}{kk^*} - \frac{2}{3kk^*}\right) > 1,$$

a contradiction. □

Bibliography

[1] Helmut Bender, *On groups with abelian Sylow 2-subgroups*, Math. Z. **117** (1970), 164–176.

[2] Walter Carlip and Wayne W. Wheeler, *Generators and Relations in the Odd Order Paper: Notes on the Junior Group Theory Seminar*, unpublished lecture notes by Walter Carlip, based on a talk by Wayne W. Wheeler on work of Thomas Peterfalvi, 1985.

[3] Charles W. Curtis and Irving Reiner, *Representation Theory of Finite Groups and Associative Algebras*, Pure and Applied Mathematics, vol. XI, Wiley-Interscience, New York, 1962.

[4] Larry L. Dornhoff, *Group Representation Theory*, vol. A, Marcel Dekker, New York, 1971.

[5] _____ , *Group Representation Theory*, vol. B, Marcel Dekker, New York, 1972.

[6] Walter Feit, *Characters of Finite Groups*, W. A. Benjamin, New York, 1967.

[7] Walter Feit, Marshall Hall Jr., and John G. Thompson, *Finite groups in which the centralizer of any non-identity element is nilpotent*, Math. Z. **74** (1960), 1–17.

[8] Walter Feit and John G. Thompson, *Solvability of groups of odd order*, Pacific J. Math. **13** (1963), 775–1029.

[9] _____ , August 20–December 31, 1991, (Personal Communication).

[10] Terence M. Gagen, *Topics in Finite Groups*, London Mathematical Society Lecture Note Series, vol. 16, Cambridge U. Press, London, 1976.

[11] George Glauberman, *A characteristic subgroup of a p-stable group*, Canad. J. Math. **20** (1968), 1101–1135.

[12] George Glauberman and Simon P. Norton, *On a combinatorial problem associated with the odd order theorem*, Proc. Amer. Math. Soc. **119** (1993), 1089–1094.

[13] David M. Goldschmidt, *A group-theoretic proof of the $p^a q^b$-theorem for odd primes*, Math. Z. **113** (1970), 373–375.

[14] Daniel Gorenstein, *Finite Groups*, second ed., Chelsea, New York, 1980.

[15] _____ , *The Classification of Finite Simple Groups*, The University Series in Mathematics, vol. 1, Plenum Press, New York, 1983.

[16] I. N. Herstein, *Topics in Algebra*, second ed., John Wiley & Sons, New York, 1975.

[17] B. Huppert, *Endliche Gruppen I*, Grundlehren der mathematischen Wissenschaften, vol. 134, Springer-Verlag, New York, 1967.

[18] B. Huppert and N. Blackburn, *Finite Groups III*, Grundlehren der mathematischen Wissenschaften, vol. 243, Springer-Verlag, New York, 1982.

[19] Marshall Hall Jr., *The Theory of Groups*, Macmillan, New York, 1959.

[20] Irving Kaplansky, *Fields and Rings*, second ed., University of Chicago Press, Chicago, Illinois, 1972.

[21] Hans Kurzweil, *Endliche Gruppen: eine Einfuhrung in die Theorie der endlichen Gruppen*, Springer-Verlag, New York, 1977.

[22] Thomas Peterfalvi, *Simplification du chapitre VI de l'article de Feit et Thompson sur les groupes d'ordre impair*, Comptes Rendus de l'Académie des Sciences **299** (1984), 531–534.

[23] William R. Scott and Fletcher Gross (eds.), *Proceedings of the Conference on Finite Groups, Park City, Utah*, New York, Academic Press, 1976.

[24] David Sibley, *Lecture notes on character theory in the odd order paper*, (unpublished).

[25] Michio Suzuki, *The nonexistence of a certain type of simple groups of odd order*, Proc. Amer. Math. Soc. **8** (1957), 686–695.

[26] _____, *Group Theory II*, Grundlehren der mathematischen Wissenschaften, vol. 248, Springer-Verlag, New York, 1986.

[27] John G. Thompson, *Factorizations of p-solvable groups*, Pacific J. Math. **16** (1966), 371–372.

List of Symbols

The page numbers indicate either the page on which the symbol is defined or the page on which the first reference to the symbol appears. In the list of symbols, p always represents a prime, P a p-group, and σ a set of primes.

FT	The original odd order paper, [8]	p. vii
CN-group	Group in which the centralizer of every nonidentity element is nilpotent	p. viii
CA-group	Group in which the centralizer of every nonidentity element is abelian	p. viii
G	Gorenstein's text, [14]	p. viii
x^α	Image of x under α	p. 1
$C_G(A)$	Set of all A-invariant elements of G	p. 1
$C_A(S)$	Set of elements of A that fix each element of S	p. 1
$[H, K]$	Subgroup generated by commutators of elements in H with elements in K	p. 1
$\langle H, K \rangle$	Subgroup generated by all elements of H and K	p. 1
$[G, A, A]$	$[[G, A], A]$	p. 1
$G^\#$	Nonidentity elements of G	p. 2
$H \lhd\lhd G$	H is a subnormal subgroup of G	p. 2
$\mathcal{O}_p(G)$	Largest normal p-subgroup of G	p. 2
$\mathcal{O}_{p'}(G)$	Largest normal p'-subgroup of G	p. 2
$\mathcal{O}_\sigma(G)$	Largest normal σ-subgroup of G	p. 2
$\mathcal{O}_{\sigma'}(G)$	Largest normal σ'-subgroup of G	p. 2
$\mathcal{O}_{p',p}(G)$	Preimage in G of $\mathcal{O}_p(G/\mathcal{O}_{p'}(G))$	p. 2
$\mathcal{O}_{p',p,p'}(G)$	Preimage in G of $\mathcal{O}_{p'}(G/\mathcal{O}_{p',p}(G))$	p. 2
Z-group	Group whose Sylow subgroups are cyclic	p. 2
$\mathscr{C}_G(T)$	$\{\, t^g \mid t \in T \text{ and } g \in G \,\}$	p. 2
TI-subgroup, (TI-subset)	Subgroup (subset) X of a group G with the property that $X \cap X^g \subseteq 1$ for all $x \in G - N_G(X)$	p. 2
$\mathcal{E}_p(G)$	Set of all elementary abelian p-subgroups of G	p. 2
$\mathcal{E}_p{}^*(G)$	Set of all maximal elementary abelian p-subgroups of G	p. 2
$\mathcal{E}_p{}^i(G)$	Set of all elementary abelian subgroups of order p^i in G	p. 2
$\mathcal{E}(G)$	Union of the sets $\mathcal{E}_p(G)$ for all primes p	p. 2

$\mathcal{E}^*(G)$	Union of the sets $\mathcal{E}_p{}^*(G)$ for all primes p	p. 2
$\mathcal{E}^i(G)$	Union of the sets $\mathcal{E}_p{}^i(G)$ for all primes p	p. 2
$Z(G)$	Center of G	p. 2
$F(G)$	Fitting subgroup of G	p. 2
$\pi(G)$	Set of primes dividing $|G|$	p. 3
$\Phi(G)$	Frattini subgroup of G	p. 5
$\Omega_1(P)$	Subgroup generated by elements of order p in P	p. 6
$E(G)$	Enveloping algebra of G	p. 9
$L\|_H, L_H$	Restriction of module L to subgroup H	p. 9
M^G	Induced $\mathbf{F}G$-module $M \otimes_{\mathbf{F}H} \mathbf{F}G$	p. 9
M^x	Conjugate module isomorphic to $M \otimes x$	p. 9
$\mathbf{GL}(V,\mathbf{F})$, $\mathbf{GL}(n,q)$	General Linear Group	p. 15
$\mathbf{SL}(V,\mathbf{F})$, $\mathbf{SL}(2,q)$	Special Linear Group	p. 15
$\mathbf{F}_p$	Finite field of p elements	p. 16
$\mathrm{m}(A)$	Minimal number of generators of A	p. 33
$\mathrm{r}_p(G), \mathrm{r}(G)$	The p-rank and rank of G, respectively	p. 33
$\mho^n(P)$	$\left\langle x^{p^n} \mid x \in P \right\rangle$	p. 33
$\mathrm{SCN}(P)$	$\{ A \mid A \lhd P \text{ and } C_P(A) = A \}$	p. 33
$\mathrm{SCN}_n(P)$	$\{ A \mid A \in \mathrm{SCN}(P) \text{ and } \mathrm{m}(A) \geq n \}$	p. 33
$G_1 \circ \cdots \circ G_n$	Central product of groups $G_1, \ldots, G_n$	p. 33
$\mathrm{cl}(G)$	Nilpotence class of G	p. 33
Z_p	Cyclic group of order p	p. 44
$Z_p \wr Z_p$	Wreath product of Z_p with Z_p	p. 44
$J(P)$	Thompson subgroup of P	p. 49
G	After p. 55, a fixed minimal counterexample to the Odd Order Theorem	p. 55
$\mathcal{M}$	Set of all maximal subgroups of G	p. 55
$\mathcal{M}(H)$	Set of maximal subgroups of G that contain H	p. 55
$\mathcal{U}$	Set of all proper subgroups H of G for which $\mathcal{M}(H)$ has a unique element	p. 55
p', π'	Set of primes other than p, or outside of the set π, respectively	p. 55
$\mathrm{SCN}_3(p)$	Set of subgroups A of G for which $A \in \mathrm{SCN}_3(P)$ for some Sylow p-subgroup P of G	p. 55
$\mathit{И}_H(A;\pi)$	Set of all π-subgroups of H normalized by A	p. 56
$\mathit{И}_H{}^*(A;\pi)$	Set of all maximal elements of $\mathit{И}_H(A;\pi)$	p. 56
$\mathit{И}_H(A;p)$	$\mathit{И}_H(A;\pi)$ for $\pi = \{p\}$	p. 56
$\mathit{И}_H{}^*(A;p)$	$\mathit{И}_H{}^*(A;\pi)$ for $\pi = \{p\}$	p. 56
π_3	Set of all primes p for which $\mathrm{r}_p(G) \geq 3$ and some Sylow p-subgroup of G normalizes some nonidentity p'-subgroup of G	p. 64
π_4	Set of all primes p for which $\mathrm{r}_p(G) \geq 3$ and $p \notin \pi_3$	p. 64
$\alpha(M)$	$\{ p \in \pi(M) \mid \mathrm{r}_p(M) \geq 3 \}$	p. 70
$\beta(M)$	$\{ p \in \alpha(M) \mid p \text{ is ideal} \}$	p. 70
$\sigma(M)$	$\{ p \in \pi(M) \mid N_G(P) \subseteq M$ for some Sylow p-subgroup P of $M \}$	p. 70
M_α	$\mathcal{O}_{\alpha(M)}(M)$	p. 70
M_β	$\mathcal{O}_{\beta(M)}(M)$	p. 70
M_σ	$\mathcal{O}_{\sigma(M)}(M)$	p. 70

$F_\sigma(M)$	$\mathcal{O}_{\sigma(M)}(F(M))$	p. 70
$F_{\sigma'}(M)$	$\mathcal{O}_{\sigma(M)'}(F(M))$	p. 70
$\tau_1(M)$	$\{\, p \in \sigma(M)' \mid p \notin \pi(M') \text{ and } \mathrm{r}_p(M) = 1 \,\}$	p. 83
$\tau_2(M)$	$\{\, p \in \sigma(M)' \mid \mathrm{r}_p(M) = 2 \,\}$	p. 83
$\tau_3(M)$	$\{\, p \in \sigma(M)' \mid p \in \pi(M') \text{ and } \mathrm{r}_p(M) = 1 \,\}$	p. 83
E_{12}	Some Hall $\tau_1(M) \cup \tau_2(M)$-subgroup of E	p. 83
E_1	Some Hall $\tau_1(M)$-subgroup of E_{12}	p. 83
E_2	Some Hall $\tau_2(M)$-subgroup of E_{12}	p. 83
E_3	Some Hall $\tau_3(M)$-subgroup of E	p. 83
σ_i	Enumeration of the sets $\sigma(M)$ for $M \in \mathcal{M}$	p. 105
$\ell_\sigma(g)$	The σ-length of g	p. 105
$\mathcal{M}_\sigma(X)$	$\{\, M \in \mathcal{M} \mid X \subseteq M_\sigma \,\}$	p. 105
$\mathcal{M}_\sigma(g)$	$\mathcal{M}_\sigma(\{g\})$	p. 105
$\kappa(M)$	$\{\, p \in \tau_1(M) \cup \tau_3(M) \mid C_{M_\sigma}(P) \neq 1$ $\text{for some } P \in \mathcal{E}_p{}^1(M) \,\}$	p. 106
$\mathcal{M}_{\mathscr{F}}$	$\{\, M \in \mathcal{M} \mid \kappa(M) \text{ is empty} \,\}$	p. 106
$\mathcal{M}_{\mathscr{P}}$	$\{\, M \in \mathcal{M} \mid \kappa(M) \text{ is not empty} \,\}$	p. 106
$\mathcal{M}_{\mathscr{P}_1}$	$\{\, M \in \mathcal{M}_{\mathscr{P}} \mid \kappa(M) = \pi(M) - \sigma(M) \,\}$	p. 106
$\mathcal{M}_{\mathscr{P}_2}$	$\{\, M \in \mathcal{M}_{\mathscr{P}} \mid \kappa(M) \neq \pi(M) - \sigma(M) \,\}$	p. 106
$R(x)$	A normal Hall subgroup of $C_G(x)$ whose action on $\mathcal{M}_\sigma(x)$ is sharply transitive	p. 108
$\widetilde{M}$	$\left\{\, xx' \mid x \in M_\sigma{}^{\#} \text{ and } x' \in R(x) \,\right\}$	p. 109
M_F	The largest normal nilpotent Hall subgroup of a maximal subgroup M of G	p. 117
M	A maximal subgroup of G	p. 117
K	A Hall $\kappa(M)$-subgroup of M	p. 117
U	A complement of KM_σ in M for which KU is a complement to M_σ in M	p. 117
K	A cyclic Hall $\kappa(M)$-subgroup of M	p. 124
Z	$K \times K^*$	p. 124
$\widehat{Z}$	$Z - (K \cup K^*)$	p. 124
$\widehat{M}_\sigma$	$\{\, a \in M \mid C_{M_\sigma}(a) \neq 1 \,\}$	p. 124
$A(M)$	$\widehat{M}_\sigma \cap UM_\sigma$	p. 124
$A_0(M)$	$\widehat{M}_\sigma - \mathscr{C}_M(K^{\#})$	p. 124
A_x	The cosets $xR(x)$ appearing in Theorem E	p. 126
$I_E(\theta)$	Stability group of the character θ	p. 133
$X \to Y$	Y is generated by abelian subgroups of G that are all normalized by X	p. 139
$L_G(X)$	The unique largest subgroup of G with the property that $X \to L_G(X)$	p. 139
$L_n(G)$	$L_G(L_{n-1}(G))$ when $n > 0$ and $\{1\}$ when $n = 0$	p. 140
$L(G)$	$\bigcap L_{2n+1}(G)$	p. 140
$L_*(G)$	$\bigcup L_{2n}(G)$	p. 140
$\mathbf{N}(a)$	Galois theoretic norm of $a \in \mathbf{F}_{p^q}$ over $\mathbf{F}_p$	p. 146
E	$\{\, a \in \mathbf{F}_{p^q} \mid \mathbf{N}(a) = \mathbf{N}(2 - a) = 1 \,\}$	p. 146
$\operatorname{card} A$	Cardinality of the set A	p. 147

Index

Page numbers in **bold** type indicate the page on which the definition or definitive reference to the indexed term appears. The symbol "G" in the index refers to a minimal counterexample to the Odd Order Theorem.

Printed in the United States
By Bookmasters